Springer Proceedings in Physics

Volume 292

Indexed by Scopus

The series Springer Proceedings in Physics, founded in 1984, is devoted to timely reports of state-of-the-art developments in physics and related sciences. Typically based on material presented at conferences, workshops and similar scientific meetings, volumes published in this series will constitute a comprehensive up to date source of reference on a field or subfield of relevance in contemporary physics. Proposals must include the following:

- Name, place and date of the scientific meeting
- A link to the committees (local organization, international advisors etc.)
- Scientific description of the meeting
- List of invited/plenary speakers
- An estimate of the planned proceedings book parameters (number of pages/articles, requested number of bulk copies, submission deadline).

Please contact:

For Americas and Europe: Dr. Zachary Evenson; zachary.evenson@springer.com
For Asia, Australia and New Zealand: Dr. Loyola DSilva; loyola.dsilva@springer.com

Giulia Ricciardi · Guglielmo De Nardo ·
Mario Merola
Editors

8th Workshop on Theory, Phenomenology and Experiments in Flavour Physics

Neutrinos, Flavor Physics and Beyond (FP@Capri2022)

Springer

Editors
Giulia Ricciardi
Department of Physics "Ettore Pancini"
University of Naples Federico II
Naples, Italy

Guglielmo De Nardo
Department of Physics "Ettore Pancini"
University of Naples Federico II
Naples, Italy

Mario Merola
University of Naples Federico II
Naples, Italy

ISSN 0930-8989 ISSN 1867-4941 (electronic)
Springer Proceedings in Physics
ISBN 978-3-031-30458-3 ISBN 978-3-031-30459-0 (eBook)
https://doi.org/10.1007/978-3-031-30459-0

This Springer imprint is published by the registered company Springer Nature Switzerland AG
The registered company address is: Gewerbestrasse 11, 6330 Cham, Switzerland

Organizing Committee and Sponsors

Organizing Committee

Prof. Giulia Ricciardi (chair),
Prof. Guglielmo De Nardo, Dipartimento di Fisica E. Pancini, Università di Napoli "Federico II" and INFN, Napoli, Italy
Prof. Mario Merola, Dipartimento di Agraria, Università di Napoli "Federico II" and INFN, Napoli, Italy

Supported by

Patrocinio del Magnifico Rettore Prof. Matteo Lorito
Università di Napoli Federico II
Dipartimento di Fisica "E. Pancini", Università di Napoli Federico II

Istituto Nazionale di Fisica Nucleare (INFN)
Sezione di Napoli

Preface

The eighth edition of the Workshop on Theory, Phenomenology and Experiments in Flavour Physics was not held in 2020 because of the coronavirus pandemic. Instead, it took place in 2022, June 11–13, at the usual venue, the International Centre for Scientific Culture of the University of Napoli Federico II, Villa Orlandi, located at Anacapri, on the island of Capri.

The 2022 edition has been a lively and successful one. About 50 invited scientists, coming from Europe, US, Korea, Japan, have actively participated; everyone seemed glad to have the opportunity to meet in presence again.

This year, the focus was on the interplay of flavour physics with neutrino and beyond the Standard Model physics. Flavour, electroweak physics and neutrino physics are all foremost in our assessment of results within the Standard Model and search for physics beyond. Anomalies in flavour physics are hints on new physics, while with neutrino masses and oscillations the new physics has already started. Analyzing interplays is fundamental to formulate a coherent framework in our ambitious quest for the ultimate laws of physics.

Arguments of debate have been, for example, the flavour anomalies, the flavour problem from leptons to quarks and back, including continuous versus discrete symmetries, flavour structure of quark and lepton in modular symmetry, the connections between the Higgs sector and neutrinos, embracing see-saw models and Higgs potential analyses. Recent results and prospects in flavour physics from ATLAS, CMS, Babar, Belle, Belle II have been presented by the Collaborations.

Seminars on neutrinos physics included global analyses, direct neutrino mass measurements, neutrino cross sections, neutrinos at high and low scales, heavy or light neutrino-antineutrino oscillations, leptogenesis, connections with dark sectors and new physics mediators, non-standard neutrino interactions, the problem of the nature of massive neutrinos, searches at colliders, neutrino telescopes, neutrinoless double beta decay experiments.

Participants found time and occasions for further discussions during the coffee break sessions in the beautiful garden of the venue, or during the social dinner at the two Michelin-starred restaurant L'Olivo of the Capri Palace Jumeirah Hotel.

It is a pleasure to express our warmest thanks to the Director of the Department of Physics E. Pancini of the University of Naples Federico II, Prof. Gennaro Miele, who has welcomed the participants in the opening address. We thank both the Department and the Istituto Nazionale di Fisica Nucleare (INFN), Sezione di Napoli, directed by Prof. Luca Lista, that granted patronage to the workshop. Heartfelt thanks are also due the Rector of the University of Naples Federico II, Prof. Matteo Lorito, for his precious encouragement and support.

Naples, Italy

Giulia Ricciardi
Chairperson

List of Speakers

Stefan Antusch, University of Basel
Marcella Bona, Queen Mary, University of London
Vedran Brdar, Fermilab and Northwestern University
Giovanni De Lellis, Istituto Nazionale di Fisica Nucleare
Patrick Foldenauer, IPPP Durham
Luigi Antonio Fusco, Università di Salerno
Jan Hajer, Técnico Lisboa, Universidade de Lisboa
Juraj Klaric, Centre for Cosmology, Particle Physics and Phenomenology (CP3), Université catholique de Louvain
Greg Landsberg, Brown University
Gaia Lanfranchi, Istituto Nazionale di Fisica Nucleare
Danny Marfatia, University of Hawaii
Ivan Martinez Soler, Harvard University
Stefano Morisi, Istituto Nazionale di Fisica Nucleare
Siavash Neshatpour, Istituto Nazionale di Fisica Nucleare
Matthias Neubert, MITP
Antimo Palano, Istituto Nazionale di Fisica Nucleare
Heinrich Päs, TU Dortmund University
Silvia Pascoli, Università di Bologna and INFN
Gil Paz, Wayne State University
Serguey Petcov, INFN/SISSA
Ezio Previtali, INFN Laboratori Nazionali del Gran Sasso
Tania Robens, Rudjer Boskovic Institute (HR)
Morimitsu Tanimoto, Niigata University
Francesco Tenchini, Università di Pisa and INFN
Arsenii Titov, University of Valencia and IFIC
Jessica Turner, Durham University
Natascia Vignaroli, Università di Napoli Federico II and INFN
Christoph Wiesinger, Technical University of Munich
Joanna Zalipska, National Centre for Nuclear Research, Warsaw

List of Participants

Stefan Antusch, University of Basel
Marcella Bona, Queen Mary, University of London
Vedran Brdar, Fermilab and Northwestern University
Eung Jin Chun, Korea Institute for Advanced Study
Giovanni De Lellis, Istituto Nazionale di Fisica Nucleare
Guglielmo De Nardo, Università di Napoli Federico II and INFN
Patrick Foldenauer, IPPP Durham
Michele Frigerio, Laboratoire Charles Coulomb, CNRS, Montpellier
Luigi Antonio Fusco, Università di Salerno
Anne Galda, Johannes Gutenberg University
Tony Gherghetta, University of Minnesota
Jan Hajer, Técnico Lisboa, Universidade de Lisboa
Ameh James, University of Jos
Sudip Jana, Max-Planck-Institut für Kernphysik
Juraj Klaric, Centre for Cosmology, Particle Physics and Phenomenology (CP3), Université catholique de Louvain
Matthias König, TU Munich
Greg Landsberg, Brown University
Gaia Lanfranchi, Istituto Nazionale di Fisica Nucleare
Danny Marfatia, University of Hawaii
Ivan Martinez Soler, Harvard University
Mario Merola, Università di Napoli Federico II and INFN
Stefano Morisi, Istituto Nazionale di Fisica Nucleare
Siavash Neshatpour, Istituto Nazionale di Fisica Nucleare
Matthias Neubert, MITP
Antimo Palano, Istituto Nazionale di Fisica Nucleare
Heinrich Päs, TU Dortmund University
Silvia Pascoli, Università di Bologna and INFN
Gil Paz, Wayne State University
Yuber F. Perez-Gonzalez, Durham University
Serguey Petcov, INFN/SISSA

Ezio Previtali, INFN Laboratori Nazionali del Gran Sasso
Giulia Ricciardi, Università di Napoli Federico II and INFN
Tania Robens, Rudjer Boskovic Institute (HR)
Ninetta Saviano, Scuola Superiore Meridionale, Università di Napoli Federico II and INFN
Manibrata Sen, Max-Planck-Institut für Kernphysik
Andrzej Szelc, University of Edinburgh
Morimitsu Tanimoto, Niigata University
Francesco Tenchini, Università di Pisa and INFN
Christoph Andreas Ternes, Istituto Nazionale di Fisica Nucleare
Arsenii Titov, University of Valencia and IFIC
Andreas Trautner, Max-Planck-Institut fuer Kernphysik Heidelberg
Jessica Turner, Durham University
Natascia Vignaroli, Università di Napoli Federico II and INFN
Christoph Wiesinger, Technical University of Munich
Felix Yu, JGU Mainz
Joanna Zalipska, National Centre for Nuclear Research, Warsaw

Contents

Contributors

Vedran Brdar Theoretical Physics Department, CERN, Esplande des Particules, Geneva 23, Switzerland

Giovanni De Lellis Dipartimento di Fisica E. Pancini, Università di Napoli Federico II, Napoli, Italy;
Istituto Nazionale di Fisica Nucleare, Napoli, Italy

Patrick Foldenauer Instituto de Física Teórica UAM-CSIC, Universidad Autónoma de Madrid, Madrid, Spain

Luigi Antonio Fusco Dipartimento di Fisica E.R. Caianiello, Università di Salerno e INFN Gruppo Collegato di Salerno, Fisciano, Italy

Jaime Hoefken Zink Dipartimento di Fisica e Astronomia, Università di Bologna, Bologna, Italy;
INFN, Sezione di Bologna, Bologna, Italy

Tobias Hurth PRISMA+ Cluster of Excellence and Institute for Physics (THEP), Johannes Gutenberg University, Mainz, Germany

Hyper-Kamiokande National Centre for Nuclear Research, Otwock, Poland

Farvah Mahmoudi CNRS/IN2P3, Institut de Physique des 2 Infinis de Lyon, Université de Lyon, Université Claude Bernard Lyon 1, Villeurbanne, France;
CERN, Theoretical Physics Department, Geneva 23, Switzerland

Diego Martinez Santos Instituto Galego de Física de Altas Enerxías, Universidade de Santiago de Compostela, Santiago de Compostela, Spain

Ivan Martinez-Soler Department of Physics and Laboratory for Particle Physics and Cosmology, Harvard University, Cambridge, MA, USA

Siavash Neshatpour INFN-Sezione di Napoli, Complesso Universitario di Monte S. Angelo, Napoli, Italy

Antimo Palano INFN, Sezione di Bari, Bari, Italy

Silvia Pascoli Dipartimento di Fisica e Astronomia, Università di Bologna, Bologna, Italy;
INFN, Sezione di Bologna, Bologna, Italy

Gil Paz Department of Physics and Astronomy, Wayne State University, Detroit, Michigan, USA

Tania Robens Rudjer Boskovic Institute, Zagreb, Croatia

Morimitsu Tanimoto Department of Physics, Niigata University, Niigata, Japan

Francesco Tenchini INFN Sezione di Pisa, Pisa, Italy;
Dipartimento di Fisica, Università di Pisa, Pisa, Italy

Arsenii Titov Departament de Física Teòrica, Universitat de València, IFIC, Universitat de València–CSIC, Burjassot, Spain

Joanna Zalipska National Centre for Nuclear Research, Otwock, Poland

Chapter 1
Recent Results from Belle and Belle II

Francesco Tenchini

Abstract B-factories provide a unique environment to test the standard model in search for new physics. The Belle and Belle II experiments succeeded each other at KEK, in Tsukuba, Japan. During its lifetime, Belle collected 1 ab^{-1} of data; at the time of this contribution, Belle II recorded 424 fb^{-1} more aiming for an unprecedented sample of 50 ab^{-1}. This unique data set will allow them to search for new physics with unmatched precision. We discuss select recent results from both experiments.

1.1 Introduction

B-factories [1] are asymmetric e^+e^- collider experiments designed to operate at or near the $\Upsilon(4S)$ resonance energy of 10.58 GeV to produce large amounts of B mesons, D mesons and τ leptons. This allows to measure the fundamental parameters of the standard model with high precision, as well as to search for deviations from its predictions which could provide a clue of new physics.

The Belle experiment was a general-purpose magnetic spectrometer located at the interaction region of one of such B-factories: the KEKB accelerator at KEK, in Tsukuba, Japan. Belle was composed of a silicon vertex detector, a 50-layer central drift chamber, an array of aerogel Cherenkov counters, a time-of-flight scintillation counter and an electromagnetic calorimeter. These detectors were located within a 1.5 T magnetic field provided by a superconducting solenoid magnet. Resistive plate chambers located outside the magnet detected K_L^0 mesons and muons. Belle is described in detail in [2]. During its operation from 1999 to 2010 it recorded

Francesco Tenchini for the Belle and Belle II collaborations.

F. Tenchini (✉)
INFN Sezione di Pisa, 56127 Pisa, Italy
e-mail: francesco.tenchini@pi.infn.it

Dipartimento di Fisica, Università di Pisa, 56127 Pisa, Italy

G. Ricciardi et al. (eds.), *8th Workshop on Theory, Phenomenology and Experiments in Flavour Physics*, Springer Proceedings in Physics 292,
https://doi.org/10.1007/978-3-031-30459-0_1

over 1 ab^{-1} of data, of which 711 fb^{-1} at the $\Upsilon(4S)$ resonance corresponding to approximately 772 million $B\overline{B}$ pairs.

Its successor Belle II is located at the same facility and paired to the SuperKEKB collider which, following the success of KEKB, was designed to reach an unprecedented instantaneous luminosity of 6×10^{35} cm^{-2}s^{-1} through the use of a novel nano-beam collision scheme [3]. The new and harsher environment in which Belle II operates demands substantial upgrades of all its subdetectors. In particular, a new tracking system was devised, composed of a two-layer silicon-pixel detector, a four-layer double-sided silicon-strip detector and a larger drift chamber, in order to provide more accurate decay vertex reconstruction closer to the interaction point.

Over its lifetime Belle II aims to record 50 ab^{-1} of data, enabling an extensive physics program [4]. At the time of this contribution Belle II recorded approximately 424 fb^{-1} of collision data before entering a technical shutdown period; 190 fb^{-1} of which were available for analysis. Although smaller than Belle's dataset, this sample can already provide significant physics input in measurements where detector performance is the dominant source of uncertainty.

In the following sections I will discuss selected recent results from both experiments.

1.2 Lifetimes of Charmed Particles

Accurate prediction of the lifetime of charmed particles is challenging, as it relies on effective models such as heavy quark expansion (HQE) to calculate strong interaction contributions at low energy. Precise measurements of these lifetimes thus provide an excellent test of these models but conversely require excellent vertex reconstruction capabilities.

The first layer of the Belle II tracking system is located only 1.4 cm from the interaction point and boasts a decay-time resolution two times better than Belle, enabling high precision absolute lifetime measurements.

Belle II reported measurements of the D^0 and D^+ lifetimes with an integrated luminosity of 72 fb^{-1} [5], and more recently of the Λ_c^+ lifetime with an integrated luminosity of 207.2 fb^{-1} [6]. These measurements were performed respectively in the $D^{*+} \to D^0(\to K^-\pi^+)\pi^+$, $D^{*+} \to D^+(\to K^-\pi^+\pi^+)\pi^0$, and $\Lambda_c^+ \to pK^-\pi^+$ decay channels and follow a common strategy. The D^{*+} and Λ_c^+ candidates are reconstructed from charged tracks identified as protons, pions and kaons, and from neutral pions reconstructed from two photons. A global decay-chain vertex fit [7] constrains each decay channel to their respective topology, and requires the D^{*+} and Λ_c^+ to originate from the interaction point. Lifetimes are extracted with a maximum likelihood fit to the unbinned distributions of the lifetime and its uncertainty (t, σ_t) (Fig. 1.1). The background contribution in the D^+ and Λ_c^+ decay-time fits is modeled from sidebands, while in the D^0 fit the background is not modeled and assigned a systematic uncertainty. An additional uncertainty is assigned to the Λ_c^+ measurement

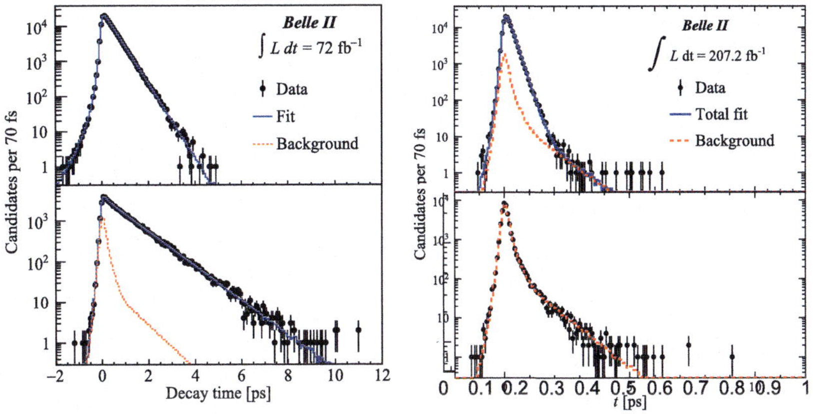

Fig. 1.1 Decay-time distributions of $D^0 \to K^-\pi^+$ (*top left*), $D^+ \to K^-\pi^+\pi^+$ (*bottom left*) and $\Lambda_c^+ \to pK^-\pi^+$ (*top right*) candidates, as well as $\Lambda_c^+ \to pK^-\pi^+$ sideband events (*bottom right*) with fit projections overlaid

due to contamination from $\Xi_c^0 \to \pi^-\Lambda_c^+$ and $\Xi_c^+ \to \pi^0\Lambda_c^+$ decays. The resulting lifetimes are

$$\tau(D^0) = 410.5 \pm 1.1(\text{stat}) \pm 0.8(\text{syst})\text{fs},$$
$$\tau(D^+) = 1030.4 \pm 4.7(\text{stat}) \pm 3.1(\text{syst})\text{fs, and}$$
$$\tau(\Lambda_c^+) = 203.20 \pm 0.89(\text{stat}) \pm 0.77(\text{syst})\text{fs},$$

which are the most precise to date and in agreement with previous results, demonstrating the vertexing capabilities of the Belle II detector and establishing the potential for future time-dependent analyses.

1.3 Hadronic B Decays

1.3.1 *Combined Measurement of ϕ_3/γ*

The CKM angle ϕ_3/γ can be measured in a theoretically clean way from the interference of $b \to c\bar{u}s$ (A_{fav}) and $b \to u\bar{c}s$ (A_{sup}) amplitudes, the latter of which is suppressed with respect to the former. Assuming the absence of new physics at tree level, this measurement of ϕ_3 can provide a standard model benchmark for other, indirect measurements which proceed at loop level. The two phases are related by

$$\frac{A_{sup}(B^- \to \overline{D}^0 K^-)}{A_{fav}(B^- \to D^0 K^-)} = r_B e^{i(\delta_B - \phi_3)} \tag{1.1}$$

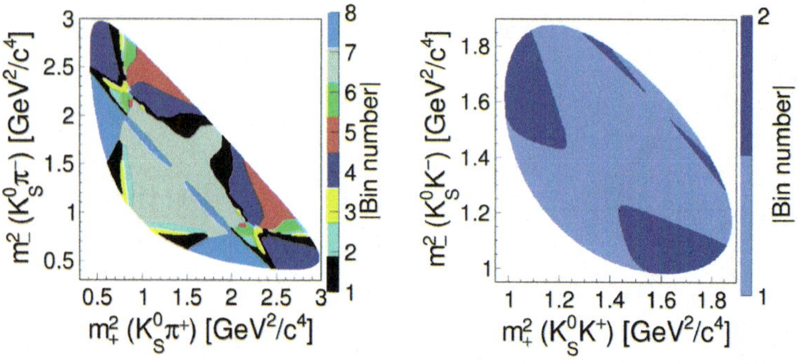

Fig. 1.2 Binning scheme used for $K_S^0\pi^+\pi^-$ (*left*) and $K_S^0K^+K^-$ (*right*) final states

where δ_B is the $D^0\overline{D}^0$ strong phase. The most precise determination of ϕ_3/γ stems from a combination of LHCb measurements [8] and has an uncertainty of $\sim 5°$. At B-factories, the most sensitive approach is based on the BPGGSZ method [9], an optimally-binned (see Fig. 1.2) Dalitz plot analysis of $B^\pm \to DK^\pm$ and $B^\pm \to D\pi^\pm$ decays with D decaying into $K_S^0\pi^+\pi^-$ and $K_S^0K^+K^-$ final states, where δ_B is constrained from external inputs. Recently Belle and Belle II published a combined analysis [10] using $711\,\mathrm{fb}^{-1}$ and $128\,\mathrm{fb}^{-1}$ of data, respectively, which significantly improves on the previous result obtained with Belle data alone. The dominant background from $e^+e^- \to q\overline{q}$ ($q = u, d, s, c$) events is characterised by different topology and suppressed by means of a multivariate classifier trained on shape variables. The yields in each Dalitz bin are then extracted with a simultaneously for $B^\pm \to DK^\pm$ and $B^\pm \to D\pi^\pm$ with a two-dimensional fit in the multivariate output and in the beam-energy difference $\Delta E = \sum_j E_j^* - E_{beam}^*$, where the sum is over every signal B decay product. The result is

$$\begin{aligned}\phi_3 &= (78.4 \pm 11.4 \pm 0.5 \pm 1.0)^\circ \\ r_B^{DK} &= 0.129 \pm 0.024 \pm 0.001 \pm 0.002 \\ \delta_B^{DK} &= (124.8 \pm 12.9 \pm 0.5 \pm 1.7)^\circ\end{aligned}$$

where the first uncertainty is statistical, the second is systematic, and the third originates from external input on the strong phase. The uncertainty on ϕ_3 improved from the 15° (statistical) and 4° (systematic) of the previous measurement thanks to better K_S^0 reconstruction, background suppression and analysis strategy. The uncertainty related to the strong-phase also improved thanks to new measurements from BESIII [11, 12]. Although the uncertainty is still larger than the current world-average value, it is also statistically limited; a future Belle II analysis with a data set corresponding to $10\,\mathrm{ab}^{-1}$ could lower it to approximately 4°.

1.3.2 Charmless Decays at Belle II

Charmless hadronic two-body decays $B \to K\pi$ provide an opportunity to test the standard model by studying direct CP violation which is sensitive to new physics contributions. In the standard model the isospin-sum rule [14]

$$\begin{aligned} \mathcal{A}_{CP}(K^+\pi^-) + \mathcal{A}_{CP}(K^0\pi^+)\frac{\mathcal{B}(K^0\pi^+)}{\mathcal{B}(K^+\pi^-)}\frac{\tau_{B^0}}{\tau_{B^+}} \\ -2\mathcal{A}_{CP}(K^+\pi^0)\frac{\mathcal{B}(K^+\pi^0)}{\mathcal{B}(K^+\pi^-)}\frac{\tau_{B^0}}{\tau_{B^+}} - 2\mathcal{A}_{CP}(K^0\pi^0)\frac{\mathcal{B}(K^0\pi^0)}{\mathcal{B}(K^+\pi^-)} = 0, \end{aligned} \tag{1.2}$$

where A_{CP} are direct CP asymmetries, holds true within 1%. The precision of this null test is dominated by experimental uncertainties on the $B \to K^0_S\pi^0$ decay mode, which can be accurately studied only at B-factories. Belle II reports a measurement of this process using 190 fb^{-1} of data [13]. B meson candidates are built using K^0_S and π^0 candidates; the flavour of the other B is determined from the remaining particles in the event using a dedicated multivariate algorithm [15]. The branching fractions and $\mathcal{A}_{CP}$ are then extracted by an extended maximum-likelihood fit of the distributions of ΔE, Δt, a modified version of the beam-constrained mass $M_{bc} = \sqrt{{E^*_{beam}}^2 - \mathbf{p}_B^2}$, and a continuum-suppressing classifier. The results are

$$\begin{aligned} \mathcal{B}(B \to K^0_S\pi^0) &= (11.0 \pm 1.2(\text{stat}) \pm 1.0(\text{syst})) \times 10^6 \\ \mathcal{A}_{CP} &= 0.41^{+0.30}_{-0.32}(\text{stat}) \pm 0.09(\text{syst}) \end{aligned}$$

in agreement with the current world average.

The radiative version of this decay, $B \to K^0_S\pi^0\gamma$, is also interesting as $b \to s\gamma$ transitions only occur at loop level in the standard model and with flavour-specific polarization. The photon is right-handed in B^0 decays and left-handed in $\overline{B}^0$ decays. Therefore, we don't expect to observe any time-dependent asymmetry. However, new physics occurring at tree level with different chirality rules could alter this picture. Belle II measured the branching fraction of this decay in preparation for a time-dependent analysis. The result, $\mathcal{B} = (7.3 \pm 1.8(\text{stat}) \pm 1.0(\text{syst})) \times 10^{-6}$, is once again in agreement with the world average value.

1.4 Semileptonic B Decays

1.4.1 Status of $|V_{ub}|$ and $|V_{cb}|$

The CKM matrix elements $|V_{ub}|$ and $|V_{cb}|$ are measured through the study of tree-level semileptonic B decays, where the standard model contribution is assumed to dominate. Such decays can be analyzed either exclusively or inclusively; in the former case a specific decay mode is reconstructed, e.g. $B \to \pi \ell \nu$, while in the latter every final state is reconstructed irrespectively of the hadronic composition: $B \to X_u \ell \nu$. The two methods, which should be in agreement, are instead subject to a long-standing 3σ discrepancy [16]. B-factories are uniquely suited to probe this puzzle by tackling the measurement in both approaches.

1.4.2 Partial Branching Fractions of $B \to X_u \ell \nu_\ell$ at Belle

Inclusive measurements of $B \to X_u \ell \nu_\ell$ are extremely challenging due to the large background from favored $B \to X_c \ell \nu_\ell$ decays, unless the measurement is performed in regions where the latter are kinematically forbidden. However, doing so introduces large theory uncertainties on the decay rate from to nonperturbative shape functions. In absence of as-of-yet-unrealized modeling improvements or model-independent approaches, it is best to extend the measurement into the $B \to X_c \ell \nu_\ell$ dominated region. Background suppression is therefore of critical importance.

To perform this measurement, Belle reconstructs the second B meson in a variety of hadronically decaying modes using a multivariate algorithm [17]. The modeling of $B \to X_c \ell \nu_\ell$ is then verified in charm-enriched sidebands; a background-suppressing classifier is trained using several discriminating observables, such as the missing squared mass or the number of observed kaons in the decay. The partial branching fraction is extracted from a template fit to the distributions of remaining events, obtaining [19]

$$\Delta\mathcal{B} = (1.59 \pm 0.07(\text{stat}) \pm 0.16(\text{syst})) \times 10^{-3}.$$

This branching fraction probes about 86% of the phase space available to the decay. It is then converted to $|V_{ub}| = (4.10 \pm 0.09 \pm 0.22 \pm 0.15) \times 10^{-3}$, where the last uncertainty denotes the theory error, using the average of several theory predictions. This result is consistent with previous inclusive determinations [20] and deviates from exclusive results by 1.3σ.

1.4.3 $B \to \pi e \nu$ at Belle II

Tagged measurements at Belle II are limited by the available sample size; nevertheless, they are important to set the framework for future analyses. Belle II produced a first measurement of the exclusive branching fractions of $B^0 \to \pi^- e^+ \nu_e$ and $B^+ \to \pi^0 e^+ \nu_e$ on $190\,\text{fb}^{-1}$ of data using a hadronic tagging approach similar to the previous section, but using a novel Full Event Interpretation algorithm [18] recently developed for Belle II.

$$\mathcal{B}(B^0 \to \pi^- e^+ \nu_e) = (1.43 \pm 0.27(\text{stat}) \pm 0.07(\text{syst})) \times 10^{-4}$$
$$\mathcal{B}(B^+ \to \pi^0 e^+ \nu_e) = (8.33 \pm 1.67(\text{stat}) \pm 0.55(\text{syst})) \times 10^{-5}$$
$$|V_{ub}| = (3.88 \pm 0.45) \times 10^{-3}$$

The results are [21] the latter of which is extracted with a combined fit using input from lattice QCD. The results agree with pre-existing exclusive determinations.

1.4.4 $|V_{cb}|$ with q^2 Moments of $B \to X_c \ell \nu$

The inclusive determination of $|V_{cb}|$ proceeds through the measurement of the differential decay rate of $B \to X_c \ell \nu$, which using HQE can be written as a power expansion of the inverse b-quark mass, $1/m_b$. The coefficients of this expansion can be directly measured on experimental data, typically from moments of the lepton energy and invariant hadronic mass, greatly reducing uncertainty from theory input. However, at higher orders of $1/m_b$ the number of these coefficients increases rapidly making the inclusion of higher power corrections increasingly difficult.

Following a recent idea [22], both Belle [23] and Belle II [24] measured the moments of the dilepton moment (q^2) spectrum for the first time. These quantities depend on a reduced set of parameters, making a fully data driven extraction of $|V_{cb}|$ possible up to $1/m_b^4$. The results from the two experiments were subsequently used in a global fit [25] to extract $|V_{cb}| = (41.69 \pm 0.59 \pm 0.23) \times 10^3$, where the first uncertainty comes from the fit and the second from higher order contributions to the power expansions. This is in good agreement with the measurement performed through the traditional approach [26] and can be further improved in the future, placing extremely strong constraints on this element of the CKM matrix.

1.5 Tests of Lepton Flavour

1.5.1 $\Omega_c^0 \to \Omega^- \ell^+ \nu_\ell$ at Belle

Electroweak coupling of gauge bosons in the standard model is assumed identical for every lepton generation; this is known as lepton flavour universality (LFU). The recent emergence of anomalies such as in $R_K = \frac{\mathcal{B}(B \to K\mu^+\mu^-)}{\mathcal{B}(B \to Ke^+e^-)}$ [27] suggest this might not be the case, which would be a clear sign of new physics.

Semileptonic decays of charmed baryons play an important role in the study of weak and strong interactions but are relatively challenging due to either low production rates or complicated backgrounds. The decay $\Omega_c^0 \to \Omega^- \ell^+ \nu_\ell$ in particular had never been tested for LFU. Belle performed the measurement [28] using a combined sample of 89.5 fb^{-1}, 711 fb^{-1} and 121.1 fb^{-1}, taken at 10.52, 10.58 and

10.86 GeVcenter-of-mass energy, respectively. The semileptonic yields are measured relatively to their hadronic counterpart, $\Omega_c^0 \to \Omega^-\pi$, by means of a binned maximum likelihood fit. They are then corrected for their relative efficiencies, resulting in the branching-fraction ratios

$$\mathcal{B}(\Omega_c^0 \to \Omega^- e^+ \nu_e)/\mathcal{B}(\Omega_c^0 \to \Omega^-\pi^+) = 1.98 \pm 0.13 \pm 0.08$$
$$\mathcal{B}(\Omega_c^0 \to \Omega^- \mu^+ \nu_\mu)/\mathcal{B}(\Omega_c^0 \to \Omega^-\pi^+) = 1.94 \pm 0.18 \pm 0.10$$
$$\mathcal{B}(\Omega_c^0 \to \Omega^- e^+ \nu_e)/\mathcal{B}(\Omega_c^0 \to \Omega^- \mu^+ \nu_\mu) = 1.02 \pm 0.10 \pm 0.02.$$

where the first uncertainty is statistical and the second systematic. The first result greatly improves on the previous measurement [29]; the second has never been measured before; and the third is also new and in good agreement with standard-model LFU predictions.

1.5.2 $\tau^\pm \to \ell^\pm \gamma$ at Belle

A related underlying assumption of the standard model is that the flavour of charged leptons is itself conserved, with immeasurably small violations becoming possible at loop level through neutrino oscillation. The same is not true in many new physics models, which envision scenarios in which lepton flavour is violated. Models predicting flavour violation between the first and the second lepton family are tightly constrained by experimental evidence, e.g. [30]; suggesting that, if indeed flavour is violated, this should occur in the third generation.

B-factories have been historically at the forefront of τ flavour violation searches, with Belle and BaBar providing most of the constraints such decays. Most recently, Belle updated their measurement of the $\tau \to \ell\gamma$ decay [31], which was previously performed on a 535 fb^{-1} data sample, to its full integrated luminosity of 988 fb^{-1}Ȧn improved selection provides better suppression of the dominant background stemming from the coincidence of $\tau \to \ell\nu\overline{\nu}$ decays with spurious photons. Belle places an upper limit to this branching fraction, at the 90% confidence level, of $\mathcal{B}(\tau^- \to e^-\gamma) < 5.6 \times 10^{-8}$ and $\mathcal{B}(\tau^- \to \mu^-\gamma) < 4.2 \times 10^{-8}$. The limit on the muon channel is the most stringent to date, superseding the previous value from BaBar [32].

1.5.3 $B^0 \to \tau^\pm \ell^\mp$ at Belle

Flavour could also be violated in tauonic B decays through loops involving NP particles such as leptoquarks [33]. This process is kinematically similar to $B^0 \to D^{(*)-}\pi^+$, which constitutes the main experimental background. Rather than explicitly reconstructing the τ decays, which can prove difficult due to the presence of neutrinos, Belle instead reconstructs [34] the other B (tag) and the light charged particle (a lepton in the signal case, a pion for the background). Energy conservation is then used to calculate the missing mass of the system, resulting in characteristic peaks at the mass of the D, D* and tau particles, with the addition of an exponential-like combinatorial background.

Signal yields are extracted with an unbinned extended maximum likelihood fit, and no signal evidence is found, resulting in an upper limit on the branching fraction, at 90% confidence level, of

$$\mathcal{B}(B^0 \to \tau^\pm e^\mp) < 1.5 \times 10^{-5}$$
$$\mathcal{B}(B^0 \to \tau^\pm \mu^\mp) < 1.6 \times 10^{-5}$$

The bound on the electron channel is the most stringent yet, while the one on the muon channel is similar to the pre-existing limit from LHCb [35].

1.6 Conclusions

B factories provide a unique environment for precision measurements. The Belle experiment, although long past its operation date, is still producing abundant physics output thanks to its large data sample. Meanwhile Belle II has entered active operation and collected so far 424 fb^{-1} of data, or approximately half that of Belle; which has already been used to provide high impact measurements. Ultimately, Belle II aims to collect 50 times the Belle data set, allowing it to search for new physics with unprecedented precision.

Acknowledgements These proceedings have received funding from the European Union's Horizon 2020 research and innovation programme under the Marie Skłodowska-Curie grant agreement No 101026516.

References

1. A.J. Bevan et al., [BaBar and Belle collaborations], Eur. Phys. J. C **74**, 3026 (2014)
2. A. Abashian et al., [Belle collaboration], Nucl. Instrum. Meth. A **479**, 117–232 (2002)
3. T. Abe et al., [Belle II collaboration]
4. E. Kou et al., [Belle II collaboration], PTEP **2019**(12), 123C01 (2019) [erratum: PTEP **2020**(2), 029201 (2020)]
5. F. Abudinén et al., [Belle II collaboration], Phys. Rev. Lett. **127**(21), 211801 (2021)
6. F. Abudinén et al., [Belle II collaboration], Phys. Rev. Lett. **130**, 071802 (2023)
7. J.F. Krohn et al., [Belle II analysis software group], Nucl. Instrum. Meth. A **976**, 164269 (2020)
8. M.W. Kenzie et al., [LHCb collaboration], LHCb-CONF-2018-002
9. A. Giri, Y. Grossman, A. Soffer, J. Zupan, Phys. Rev. D **68**, 054018 (2003)
10. F. Abudinén et al., [Belle and Belle II collaborations], JHEP **02**, 063 (2022)
11. M. Ablikim et al., [BESIII collaboration], Phys. Rev. D **101**(11), 112002 (2020)

12. M. Ablikim et al., [BESIII collaboration] Phys. Rev. D **102**, 052008 (2020)
13. F. Abudinén et al., [Belle II collaboration], arXiv:2206.07453 [hep-ex]
14. M. Gronau, Phys. Lett. B **627**, 82–88 (2005)
15. F. Abudinén et al., [Belle II collaboration] Eur. Phys. J. C **82**(4), 283 (2022)
16. Y.S. Amhis et al., [HFLAV], Eur. Phys. J. C **81**(3), 226 (2021)
17. M. Feindt, F. Keller, M. Kreps, T. Kuhr, S. Neubauer, D. Zander, A. Zupanc, Nucl. Instrum. Meth. A **654**, 432–440 (2011)
18. T. Keck, F. Abudinén, F.U. Bernlochner, R. Cheaib, S. Cunliffe, M. Feindt, T. Ferber, M. Gelb, J. Gemmle, P. Goldenzweig et al., Comput. Softw. Big Sci. **3**(1), 6 (2019)
19. L. Cao et al., [Belle collaboration], Phys. Rev. D **104**(1), 012008 (2021)
20. J.P. Lees et al., [BaBar collaboration], Phys. Rev. D **95**(7), 072001 (2017)
21. F. Abudinén et al., [Belle II collaboration], arXiv:2206.08102 [hep-ex]
22. M. Fael, T. Mannel, K. Keri Vos, JHEP **02**, 177 (2019)
23. R. van Tonder et al., [Belle collaboration], Phys. Rev. D **104**(11), 112011 (2021)
24. F. Abudinén et al. [Belle II Collaboration], Phys. Rev. D **107**, 072002
25. F. Bernlochner, M. Fael, K. Olschewsky, E. Persson, R. van Tonder, K.K. Vos, M. Welsch, JHEP **10**, 068 (2022)
26. M. Bordone, B. Capdevila, P. Gambino, Phys. Lett. B **822**, 136679 (2021)
27. R. Aaij et al., [LHCb collaboration], Nature Phys. **18**(3), 277–282 (2022)
28. Y.B. Li et al., [Belle collaboration], Phys. Rev. D **105**(9), L091101 (2022)
29. R. Ammar et al., [CLEO collaboration], Phys. Rev. Lett. **89**, 171803 (2002)
30. A.M. Baldini et al., [MEG collaboration], Eur. Phys. J. C **76**(8), 434 (2016)
31. A. Abdesselam et al., [Belle collaboration], JHEP **10**, 19 (2021)
32. B. Aubert et al., [BaBar collaboration], Phys. Rev. Lett. **104**, 021802 (2010)
33. A.D. Smirnov, Mod. Phys. Lett. A **33**, 1850019 (2018)
34. H. Atmacan et al., [Belle collaboration], Phys. Rev. D **104**(9), L091105 (2021)
35. R. Aaij et al., [LHCb collaboration], Phys. Rev. Lett. **123**(21), 211801 (2019)

Chapter 2
Neutral Current B-Decay Anomalies

Siavash Neshatpour, Tobias Hurth, Farvah Mahmoudi, and Diego Martinez Santos

Abstract We discuss the implications of $b \to s\ell^+\ell^-$ measurements and their deviations with respect to the Standard Model predictions in a model-independent framework. We highlight in particular the impact of the recent updated measurements including the updated $B_s \to \phi\mu^+\mu^-$ branching ratios and angular observables, the recent CMS measurement of the branching ratio of $B_s \to \mu^+\mu^-$, and the LHCb measured lepton flavour universality violating ratios $R_{K_S^0}$ and $R_{K^{*+}}$. In addition, we check the compatibility of the new physics effect for the theoretically clean observables with the rest of the neutral B decays observables.

CERN-TH-2022-166, MITP-22-081.

S. Neshatpour (✉)
INFN-Sezione di Napoli, Complesso Universitario di Monte S. Angelo, Via Cintia Edificio 6, 80126 Napoli, Italy
e-mail: neshatpour@na.infn.it

T. Hurth
PRISMA+ Cluster of Excellence and Institute for Physics (THEP),
Johannes Gutenberg University, D-55099 Mainz, Germany
e-mail: Tobias.Hurth@cern.ch

F. Mahmoudi
CNRS/IN2P3, Institut de Physique des 2 Infinis de Lyon,break Université de Lyon, Université Claude Bernard Lyon 1, UMR 5822, 69622 Villeurbanne, France
e-mail: nazila@cern.ch

CERN, Theoretical Physics Department, CH-1211 Geneva 23, Switzerland

D. Martinez Santos
Instituto Galego de Física de Altas Enerxías, Universidade de Santiago de Compostela, Santiago de Compostela, Spain
e-mail: Diego.Martinez.Santos@cern.ch

G. Ricciardi et al. (eds.), *8th Workshop on Theory, Phenomenology and Experiments in Flavour Physics*, Springer Proceedings in Physics 292,
https://doi.org/10.1007/978-3-031-30459-0_2

2.1 Introduction

In the last few years, since the measured deviation in the angular observable P_5' of the $B \to K^*\mu^+\mu^-$ decay [1], there have been several measurements in neutral B-decays indicating tension with the Standard Model (SM). Updated measurements by LHCb for the $P_5'(B \to K^*\mu^+\mu^-)$ have persistently shown tension with the SM which can be explained with short distance new physics (NP) contributions [2, 3]. This is also the case of the overall $B \to K^*\mu^+\mu^-$ angular observables and is supported in addition (see e.g. [4]) by the angular analysis of its isospin partner in the recent measurement of $B^+ \to K^{*+}\mu^+\mu^-$ [5]. The $B_s \to \phi\mu^+\mu^-$ branching fraction [6–8] also indicates tensions with the SM and is measured to be below the SM prediction. This trend is seen in several other $b \to s\ell^+\ell^-$ branching fractions such as $B \to K\mu^+\mu^-$ [9] and $\Lambda_b \to \Lambda\mu^+\mu^-$ [10]. Since the branching fractions are dependent on the relevant local form factors, they suffer from large theoretical uncertainties. In contrast, the angular observables have a reduced sensitivity to the form factor uncertainties, but they are still dependent on the non-local hadronic contributions whose size are not fully known in QCD factorisation. Consequently the significance of the anomalies are dependent on the estimated size of the non-local effects. Recent theoretical progress for a better control of these effects can be found in [11–13].

A set of observables to test lepton flavour universality violation (LFUV) in $b \to s\ell^+\ell^-$ transitions is defined as $R_H = (B \to H\mu^+\mu^-)/(B \to He^+e^-)$ with $H = K^+, K^*, \phi, ...$ [14]. Unlike the observables mentioned in the previous paragraph, these ratios are very precisely known in the SM. There have been signs of deviation from the SM in the LFUV ratios for the case of R_K [15–17] and R_{K^*} [18]. The recent measurements of $R_{K_S^0}$ and $R_{K^{*+}}$ [19] although within 2σ of the SM prediction, show the same trend as their isospin partners with the central values below the SM predictions. Incidentally, there have also been a slight sign of LFUV in flavour changing neutral current processes in the Kaon sector [20] (currently the experimental uncertainty is quite large for these processes).

The significance of each of the B-anomalies, individually is around $\sim 2-3\sigma$, however collectively they can be explained by common NP scenarios and have a much larger significance in a global analysis [21–26].

Another precisely predicted observable with an uncertainty of less than 5% in its SM prediction is BR($B_s \to \mu^+\mu^-$) which has been measured by several experiments. Previous measurements of ATLAS [27], CMS [28] and LHCb [29, 30] were in about 2σ tension with the SM prediction [21]. However, the situation has changed with the recent data from CMS [31]. For our fits, we combined the ATLAS [27] and LHCb [29, 30] results together with the recent CMS [31] measurement, considering a joint 2D likelihood as shown in Fig. 2.1. We obtained the experimental combined value of the $B_s \to \mu^+\mu^-$ branching ratio to be

$$\mathrm{BR}(\mathrm{B_s} \to \mu^+\mu^-)^{\mathrm{comb.}}_{\mathrm{exp}} = \left(3.52^{+0.32}_{-0.30}\right) \times 10^{-9}\,, \tag{2.1}$$

which is within 1σ of the SM prediction.

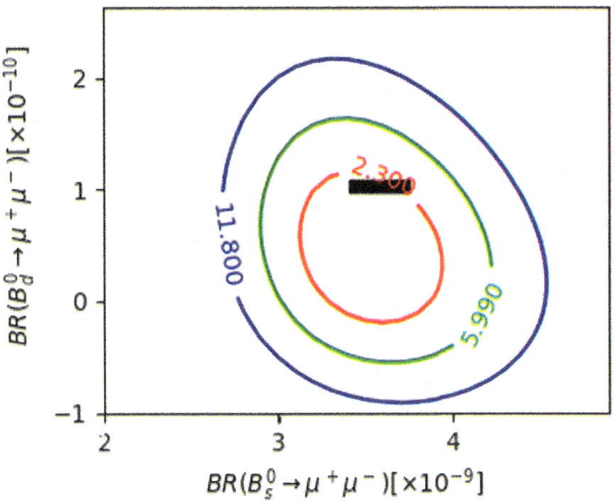

Fig. 2.1 Two dimensional likelihood plot of $\text{BR}(B_{s,d} \to \mu^+\mu^-)$

2.2 Coherence of Clean Observables with the Rest of the Rare *B*-Decay Observables

In order to examine the consistency of the implication of the clean observables for new physics as compared to the rest of the observables [32, 33], we perform two sets of fits; one to the clean observables where we consider R_K, R_{K^*} as well as their isospin partners $R_{K_S^0}$ and $R_{K^{*+}}$ [19] and also $\text{BR}(B_{s,d} \to \mu^+\mu^-)$, and another one considering the rest of the $bs\ell\ell$ observables. The observable calculations and the χ^2 fitting is done using the SuperIso public program [34–38].

2.2.1 *Clean Observables*

In Table 2.1 we give the one-dimensional NP fits to clean observables and compare them with our 2021 fit results [21]. Compared to [21], we now include the two LFUV ratios $R_{K_S^0}$ and $R_{K^{*+}}$ [19] as well as the R_K measurement by Belle [39] in the [1, 6] GeV^2 bin and the updated combination for $\text{BR}(B_s \to \mu^+\mu^-)$ as given in (2.1). While the significance of NP in $C_9^{e,\mu}$ or C_{10}^e has slightly increased, the C_{10}^μ solution is now less favoured compared to the 2021 results [21]. This is expected as the new combination of $\text{BR}(B_s \to \mu^+\mu^-)$ is now in much better agreement with the SM prediction and constrains more C_{10}^μ. The inclusion of $\text{BR}(B_s \to \mu^+\mu^-)$ in this set of observables is crucial in breaking the degeneracy between NP in δC_9^μ and δC_{10}^μ for explaining the measured values of the LFUV ratios as can be clearly seen in Fig. 2.2 where without $\text{BR}(B_s \to \mu^+\mu^-)$ the best fit point of $R_{K^{(*)}}$ is given by the

Table 2.1 Comparison of the fits to clean observables with the 2021 fit results [21] on the left and the updated 2022 fits on the right

Only LFUV ratios and $B_{s,d} \to \ell^+\ell^-$ **2021 fit results** $(\chi^2_{\rm SM} = 28.19)$				Only LFUV ratios and $B_{s,d} \to \ell^+\ell^-$ **2022 fit results** $(\chi^2_{\rm SM} = 30.63)$			
	b.f. value	$\chi^2_{\rm min}$	Pull$_{\rm SM}$		b.f. value	$\chi^2_{\rm min}$	Pull$_{\rm SM}$
δC_9	-1.00 ± 6.00	28.1	0.2σ	δC_9	-2.00 ± 5.00	30.5	0.4σ
δC_9^e	0.80 ± 0.21	11.2	4.1σ	δC_9^e	0.83 ± 0.21	10.8	4.4σ
δC_9^μ	-0.77 ± 0.21	11.9	4.0σ	δC_9^μ	-0.80 ± 0.21	11.8	4.3σ
δC_{10}	0.43 ± 0.24	24.6	1.9σ	δC_{10}	0.03 ± 0.20	30.6	0.1σ
δC_{10}^e	-0.78 ± 0.20	9.5	4.3σ	δC_{10}^e	-0.81 ± 0.19	8.7	4.7σ
δC_{10}^μ	0.64 ± 0.15	7.3	4.6σ	δC_{10}^μ	0.50 ± 0.14	16.2	3.8σ
$\delta C_{\rm LL}^e$	0.41 ± 0.11	10.3	4.2σ	$\delta C_{\rm LL}^e$	0.43 ± 0.11	9.7	4.6σ
$\delta C_{\rm LL}^\mu$	-0.38 ± 0.09	7.1	4.6σ	$\delta C_{\rm LL}^\mu$	-0.33 ± 0.08	12.4	4.3σ

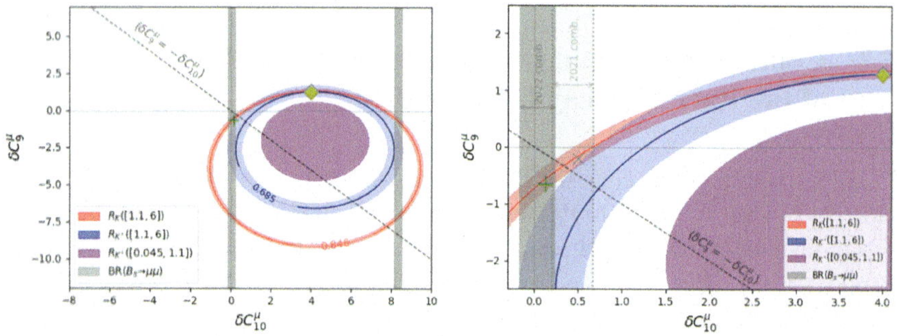

Fig. 2.2 The prediction of $R_{K^{(*)}}$ and $\mathrm{BR}(B_s \to \mu^+\mu^-)$ within 1σ $\left(= \sqrt{\sigma^2_{\rm th} + \sigma^2_{\rm exp}}\right)$ of their measured values. On the right plot we have the zoomed-in version of the left plot. The dark gray band indicates the 1σ region corresponding to the updated combination of $\mathrm{BR}(B_s \to \mu\mu)$ and the lighter gray region (on the right plot) with the dotted borders corresponds to the 2021 combination. The yellow diamond indicates the best fit value to $R_{K^{(*)}}$, while the green plus sign (gray cross) corresponds to the best fit point when the 2022 (2021) combination for $\mathrm{BR}(B_s \to \mu\mu)$ is included in the fit

yellow diamond while including it moves the best fit value to the green plus sign. The impact of the updated value of the $\mathrm{BR}(B_s \to \mu^+\mu^-)$ can bee seen in the right plot by comparing the green plus sign with the gray cross corresponding to the best fit point when the 2021 combination for $\mathrm{BR}(B_s \to \mu^+\mu^-)$ was considered.

2.2.2 All Except the Clean Observables

We consider now the 1-dimensional NP fits to the rest of the observables, excluding the LFUV ratios and $B_{s,d} \to \ell^+\ell^-$. We assume 10% power correction for the non-factorisable contributions beyond QDC factorisation [40–42]. Compared to [21] we use the updated LHCb results for the $B_s \to \phi\mu^+\mu^-$ observables [7, 8] with 8.4 fb^{-1} of data. The CMS measurement for $F_H(B^+ \to K^+\mu^+\mu^-)$ [43] and the LHCb measurement of the angular observables of $B \to K^* e^+e^-$ [44] have also been considered. As can be seen in Table 2.2, the hierarchy of the preferred NP contributions is similar to the 2021 results, where the most preferred scenarios are still NP in lepton flavour violating δC_9^μ and NP in lepton flavour universal δC_9 with the third most preferred description given by NP in the chiral basis δC_{LL}^μ. The above mentioned scenarios however are showing a $\sim 0.4\sigma$ reduced significance compared to our 2021 results which is mainly due to the updated $B_s \to \phi\mu^+\mu^-$ experimental data.

For the fit to all observables except the clean ones there is no significant indication for NP within the electron sector since not only the measurements in the electron sector are in good agreement with their SM predictions, there are also far less data compared to the decays with muons. Comparing the result of Table 2.2 with the result of the previous subsection (Table 2.1) we see that there is not a full agreement for the preferred scenarios, however, there are common scenarios such as NP contributions to δC_9^μ [45, 46] which have a large significance for both datasets with best fit points that agree within 1σ.

The compatibility of the two-dimensional NP fits to "clean observables" and the NP fits to "all observables except the clean ones" can be seen in Fig. 2.3 where also the significant impact of including or removing $\text{BR}(B_s \to \mu^+\mu^-)$ from each dataset is clearly visible, especially for the clean observables.

Table 2.2 Comparison of the fits to all observables except the clean ones with the 2021 fit results on the left and the updated 2022 fits on the right

All observables except LFUV ratios and $B_{s,d} \to \ell^+\ell^-$ **2021 fit results** ($\chi^2_{\text{SM}} = 200.1$)				All observables except LFUV ratios and $B_{s,d} \to \ell^+\ell^-$ **2022 fit results** ($\chi^2_{\text{SM}} = 221.8$)			
	b.f. value	χ^2_{min}	Pull_{SM}		b.f. value	χ^2_{min}	Pull_{SM}
δC_9	-1.01 ± 0.13	158.2	6.5σ	δC_9	-0.95 ± 0.13	185.1	6.1σ
δC_9^e	0.70 ± 0.60	198.8	1.1σ	δC_9^e	0.70 ± 0.60	220.5	1.1σ
δC_9^μ	-1.03 ± 0.13	156.0	6.6σ	δC_9^μ	-0.96 ± 0.13	182.8	6.2σ
δC_{10}	0.34 ± 0.23	197.7	1.5σ	δC_{10}	0.29 ± 0.21	219.8	1.4σ
δC_{10}^e	-0.50 ± 0.50	199.0	1.0σ	δC_{10}^e	-0.60 ± 0.50	220.6	1.1σ
δC_{10}^μ	0.41 ± 0.23	196.5	1.9σ	δC_{10}^μ	0.35 ± 0.20	218.7	1.8σ
δC_{LL}^e	0.33 ± 0.29	198.9	1.1σ	δC_{LL}^e	0.34 ± 0.29	220.6	1.1σ
δC_{LL}^μ	-0.75 ± 0.13	167.9	5.7σ	δC_{LL}^μ	-0.64 ± 0.13	195.0	5.2σ

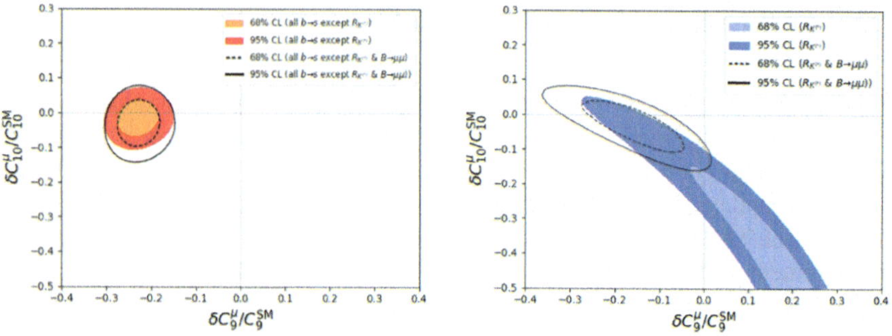

Fig. 2.3 Two-dimensional fits to the clean observables on the right and to the rest of the observables on the left

2.3 Global Fits to All $b \to s\ell^+\ell^-$ Observables

For a global analysis of the NP implications of rare B-decays, we need to take into account all relevant $b \to s$ decays combining the datasets of Sects. 2.2.1 and 2.2.2. We assume 10% error for the power corrections when applicable.

2.3.1 One- and Two-Dimensional Fits

The 1-dim NP fits to the rare B-decays are given in Table 2.3. As anticipated from the comparison of the fits to clean observables and the rest of the observables, the most favoured scenario to explain the overall data is lepton flavour violating NP in δC_9^{μ}. The other prominent scenarios are NP in $\delta C_{\rm LL}^{\mu}$ followed by lepton flavour universal NP in δC_9. While the hierarchy of the favoured scenarios has not changed, it should be noted that NP in δC_{10}^{μ} is now less favoured which is mostly due to the updated combination for $\text{BR}(B_s \to \mu^+\mu^-)$. This can also be seen in the decrease of the significance of δC_{LL}^{μ} compared to the 2021 fit results.

The decrease of preference of NP in δC_{10}^{μ} can also be seen in the 2-dim fits of Fig. 2.4. In the 1- and 2-dim fits of Table 2.3 and Fig. 2.4 we have not shown the NP fits to the radiative coefficient δC_7, the scalar and pseudoscalar coefficients ($\delta C_{Q_{1,2}}$) or the coefficients where the hadronic currents are right-handed ($\delta C_i'$) since they are all strongly constrained by data. The situation can in principle change when several coefficients can simultaneously contribute, this is clearly the case when doing a simultaneous fit to δC_{10}^{μ} and $\delta C_{Q_{1,2}}^{\mu}$ [47] which would otherwise be severely constrained by $\text{BR}(B_s \to \mu^+\mu^-)$ if only one single coefficient would contribute.

Table 2.3 Comparison of the fits to all observables with the 2021 fit results on the left and the updated 2022 fits on the right

All observables **2021 fit results** ($\chi^2_{\text{SM}} = 225.8$)				All observables **2022 fit results** ($\chi^2_{\text{SM}} = 253.5$)			
	b.f. value	χ^2_{min}	Pull$_{\text{SM}}$		b.f. value	χ^2_{min}	Pull$_{\text{SM}}$
δC_9	-0.99 ± 0.13	186.2	6.3σ	δC_9	-0.95 ± 0.13	215.8	6.1σ
δC_9^e	0.79 ± 0.20	207.7	4.3σ	δC_9^e	0.82 ± 0.19	232.4	4.6σ
δC_9^μ	-0.95 ± 0.12	168.6	7.6σ	δC_9^μ	-0.92 ± 0.11	195.2	7.6σ
δC_{10}	0.32 ± 0.18	222.3	1.9σ	δC_{10}	0.08 ± 0.16	253.2	0.5σ
δC_{10}^e	-0.74 ± 0.18	206.3	4.4σ	δC_{10}^e	-0.77 ± 0.18	230.6	4.8σ
δC_{10}^μ	0.55 ± 0.13	205.2	4.5σ	δC_{10}^μ	0.43 ± 0.12	238.9	3.8σ
δC_{LL}^e	0.40 ± 0.10	206.9	4.3σ	δC_{LL}^e	0.42 ± 0.10	231.4	4.7σ
δC_{LL}^μ	-0.49 ± 0.08	180.5	6.7σ	δC_{LL}^μ	-0.43 ± 0.07	213.6	6.3σ

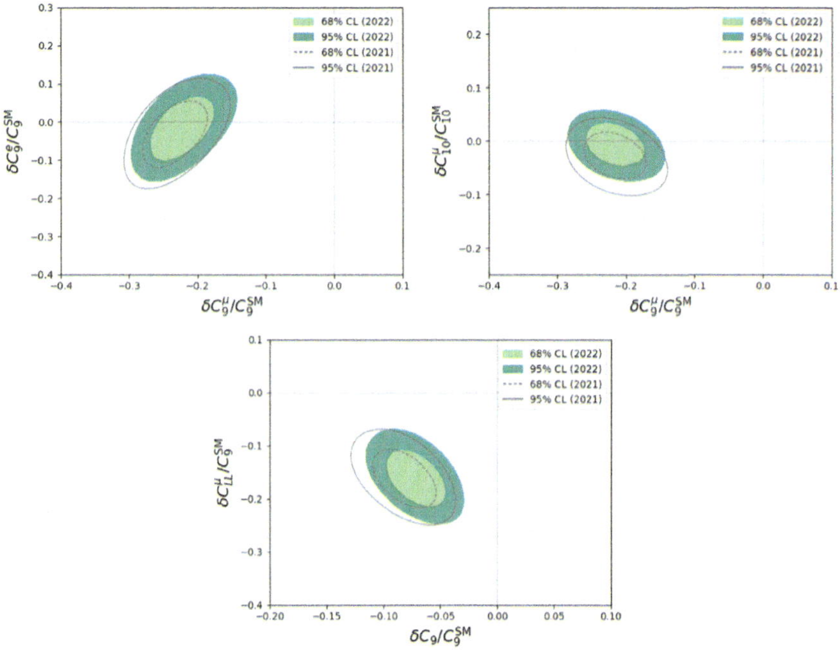

Fig. 2.4 Two-dimensional fit to all rare B-decay observables

2.3.2 Multidimensional Fit

A multidimensional fit gives in principle a more realistic picture than assuming new physics contribution to only a single coefficient, as it is very unlikely for a UV-complete scenario to merely affect one coefficient while the rest of the coefficients are kept to their SM values. Therefore, here we consider a 20-dim fit varying all relevant Wilson coefficients (Table 2.4). Besides being more realistic, this multidimensional fit has the advantage of avoiding the look elsewhere effect (LEE) since LEE not only takes place when one makes a selected choice of observables but is also relevant in the case when a posteriori a subset of specific NP directions are assumed which is circumvented when all possible Wilson coefficients are varied. With a large set of free parameters and the limited decay modes there can be flat directions or non-sensitive NP coefficients that can be removed by considering the correlations and likelihood profiles in order to get an "effective" number of degrees of freedom ($\mathrm{dof_{eff}}$). In the 20-dim fit we find degeneracy in δC_{10}^{e} and $\delta C_{10}^{e\prime}$ which results in having $\mathrm{dof_{eff}} = 19$. With the current data, there are still several of the Wilson coefficients which are only loosely constrained, especially in the electron sector where there is less data. The significance of the NP in our 20-dim fit is 5.5σ, remaining the same as what we had found in [21].

Table 2.4 Comparison of 20-dim fit to all observables with the 2022 (2021) result on the right (left). The $\mathrm{Pull_{SM}}$ in the parenthesis is given for $\mathrm{dof_{eff}} = 19$

All observables with $\chi^2_{\rm SM} = 225.8$, nr. obs.= 173 **2021 fit results** ($\chi^2_{\rm min} = 151.6$; $\mathrm{Pull_{SM}} = 5.5(5.6)\sigma$)				All observables with $\chi^2_{\rm SM} = 253.5$, nr. obs.= 183 **2022 fit results** ($\chi^2_{\rm min} = 179.1$; $\mathrm{Pull_{SM}} = 5.5(5.5)\sigma$)			
δC_7 0.05 ± 0.03		δC_8 -0.70 ± 0.40		δC_7 0.06 ± 0.03		δC_8 -0.80 ± 0.40	
$\delta C_7'$ -0.01 ± 0.02		$\delta C_8'$ 0.00 ± 0.80		$\delta C_7'$ -0.01 ± 0.01		$\delta C_8'$ -0.30 ± 1.30	
δC_9^{μ} $-1.16 \pm$ 0.17	δC_9^{e} $-6.70 \pm$ 1.20	δC_{10}^{μ} $0.20 \pm$ 0.21	δC_{10}^{e} degenerate w/↓	δC_9^{μ} $-1.14 \pm$ 0.19	δC_9^{e} $-6.50 \pm$ 1.90	δC_{10}^{μ} $0.21 \pm$ 0.20	δC_{10}^{e} degenerate w/↓
$\delta C_9^{\prime\mu}$ $0.09 \pm$ 0.34	$\delta C_9^{\prime e}$ $1.90 \pm$ 1.50	$\delta C_{10}^{\prime\mu}$ $-0.12 \pm$ 0.20	$\delta C_{10}^{\prime e}$ degenerate w/↑	$\delta C_9^{\prime\mu}$ $0.05 \pm$ 0.32	$\delta C_9^{\prime e}$ $1.40 \pm$ 2.30	$\delta C_{10}^{\prime\mu}$ $-0.03 \pm$ 0.19	$\delta C_{10}^{\prime e}$ degenerate w/↑
$\delta C_{Q_1}^{\mu}$ $0.04 \pm$ 0.10	$\delta C_{Q_1}^{e}$ $-1.50 \pm$ 1.50	$\delta C_{Q_2}^{\mu}$ $-0.09 \pm$ 0.10	$\delta C_{Q_2}^{e}$ $-4.10 \pm$ 1.5	$\delta C_{Q_1}^{\mu}$ $0.04 \pm$ 0.20	$\delta C_{Q_1}^{e}$ $-1.60 \pm$ 1.70	$\delta C_{Q_2}^{\mu}$ $-0.15 \pm$ 0.08	$\delta C_{Q_2}^{e}$ $-4.10 \pm$ 0.9
$\delta C_{Q_1}^{\prime\mu}$ $0.15 \pm$ 0.10	$\delta C_{Q_1}^{\prime e}$ $-1.70 \pm$ 1.20	$\delta C_{Q_2}^{\prime\mu}$ $-0.14 \pm$ 0.11	$\delta C_{Q_2}^{\prime e}$ $-4.20 \pm$ 1.2	$\delta C_{Q_1}^{\prime\mu}$ $-0.03 \pm$ 0.20	$\delta C_{Q_1}^{\prime e}$ $-1.50 \pm$ 2.10	$\delta C_{Q_2}^{\prime\mu}$ $-0.16 \pm$ 0.08	$\delta C_{Q_2}^{\prime e}$ $-4.00 \pm$ 1.2

2.4 Conclusions

We presented the NP fits to rare B decays which include the recent measurements of $B_s \to \phi\mu^+\mu^-$ observables and the lepton flavour violating ratios $R_{K^{*+}}$ and R_{K_S} by LHCb. We have furthermore updated the BR($B_s \to \mu^+\mu^-$) combination to include the very recent measurement by CMS. The main change in the NP fits is a reduction for the significance of a δC_{10}^{μ} solution or the scenarios involving it which is mainly due to the recent BR($B_s \to \mu^+\mu^-$) measurement being in agreement with the SM value. However, the hierarchy of the favoured scenarios for the global fit has remained stable and the preferred scenario is still NP with δC_9^{μ}. We also updated our twenty dimensional fit which avoids the look elsewhere effect finding a 5.5σ significance.

Acknowledgements SN is grateful to FPCapri2022 organisers for the invitation to present this work.

References

1. LHCb collaboration, Measurement of form-factor-independent observables in the decay $B^0 \to K^{*0}\mu^+\mu^-$. Phys. Rev. Lett. **111**, 191801 (2013). https://doi.org/10.1103/PhysRevLett.111.191801. arXiv:1308.1707
2. LHCb collaboration, Angular analysis of the $B^0 \to K^{*0}\mu^+\mu^-$ decay using 3 fb^{-1} of integrated luminosity. JHEP **02**, 104 (2016). https://doi.org/10.1007/JHEP02(2016)104. arXiv:1512.04442
3. LHCb collaboration, Measurement of CP-averaged observables in the $B^0 \to K^{*0}\mu^+\mu^-$ decay. Phys. Rev. Lett. **125**, 011802 (2020). https://doi.org/10.1103/PhysRevLett.125.011802. arXiv:2003.04831
4. T. Hurth, F. Mahmoudi, S. Neshatpour, Model independent analysis of the angular observables in $B^0 \to K^{*0}\mu^+\mu^-$ and $B^+ \to K^{*+}\mu^+\mu^-$. Phys. Rev. D **103**, 095020 (2021). https://doi.org/10.1103/PhysRevD.103.095020. arXiv:2012.12207
5. LHCb collaboration, Angular analysis of the $B^+ \to K^{*+}\mu^+\mu^-$ decay. Phys. Rev. Lett. **126**, 161802 (2021). https://doi.org/10.1103/PhysRevLett.126.161802. arXiv:2012.13241
6. LHCb collaboration, Angular analysis and differential branching fraction of the decay $B_s^0 \to \phi\mu^+\mu^-$. JHEP **09**, 179 (2015). https://doi.org/10.1007/JHEP09(2015)179. arXiv:1506.08777
7. LHCb collaboration, Angular analysis of the rare decay $B_s^0 \to \phi\mu^+\mu^-$. JHEP **11**, 043 (2021). https://doi.org/10.1007/JHEP11(2021)043. arXiv:2107.13428
8. LHCb collaboration, Branching fraction measurements of the rare $B_s^0 \to \phi\mu^+\mu^-$ and $B_s^0 \to f_2'(1525)\mu^+\mu^-$- decays. Phys. Rev. Lett. **127**, 151801 (2021). https://doi.org/10.1103/PhysRevLett.127.151801. arXiv:2105.14007
9. LHCb collaboration, Differential branching fractions and isospin asymmetries of $B \to K^{(*)}\mu^+\mu^-$ decays. JHEP **06**, 133 (2014). https://doi.org/10.1007/JHEP06(2014)133. arXiv:1403.8044
10. LHCb collaboration, Differential branching fraction and angular analysis of $\Lambda_b^0 \to \Lambda\mu^+\mu^-$ decays. JHEP **06**, 115 (2015). https://doi.org/10.1007/JHEP06(2015)115. arXiv:1503.07138
11. C. Bobeth, M. Chrzaszcz, D. van Dyk, J. Virto, Long-distance effects in $B \to K^*\ell\ell$ from analyticity. Eur. Phys. J. C **78**, 451 (2018). https://doi.org/10.1140/epjc/s10052-018-5918-6. arXiv:1707.07305
12. N. Gubernari, D. van Dyk, J. Virto, Non-local matrix elements in $B_{(s)} \to \{K^{(*)}, \phi\}\ell^+\ell^-$. JHEP **02**, 088 (2021). https://doi.org/10.1007/JHEP02(2021)088. arXiv:2011.09813

13. N. Gubernari, M. Reboud, D. van Dyk, J. Virto, Improved theory predictions and global analysis of exclusive $b \to s\mu^+\mu^-$ processes. arXiv:2206.03797
14. G. Hiller, F. Kruger, More model-independent analysis of $b \to s$ processes. Phys. Rev. D **69**, 074020 (2004). https://doi.org/10.1103/PhysRevD.69.074020. arXiv:hep-ph/0310219
15. LHCb collaboration, Test of lepton universality using $B^+ \to K^+\ell^+\ell^-$ decays. Phys. Rev. Lett. **113**, 151601 (2014). https://doi.org/10.1103/PhysRevLett.113.151601. arXiv:1406.6482
16. LHCb collaboration, Search for lepton-universality violation in $B^+ \to K^+\ell^+\ell^-$ decays. Phys. Rev. Lett. **122**, 191801 (2019). https://doi.org/10.1103/PhysRevLett.122.191801. arXiv:1903.09252
17. LHCb collaboration, Test of lepton universality in beauty-quark decays. Nature Phys. **18**, 277 (2022). https://doi.org/10.1038/s41567-021-01478-8. arXiv:2103.11769
18. LHCb collaboration, Test of lepton universality with $B^0 \to K^{*0}\ell^+\ell^-$ decays. JHEP **08**, 055 (2017). https://doi.org/10.1007/JHEP08(2017)055. arXiv:1705.05802
19. LHCb collaboration, Tests of lepton universality using $B^0 \to K_S^0\ell^+\ell^-$ and $B^+ \to K^{*+}\ell^+\ell^-$ decays. Phys. Rev. Lett. **128**, 191802 (2022). https://doi.org/10.1103/PhysRevLett.128.191802. arXiv:2110.09501
20. G. D'Ambrosio, A.M. Iyer, F. Mahmoudi, S. Neshatpour, Anatomy of kaon decays and prospects for lepton flavour universality violation. JHEP **09**, 148 (2022). https://doi.org/10.1007/JHEP09(2022)148. arXiv:2206.14748
21. T. Hurth, F. Mahmoudi, D.M. Santos, S. Neshatpour, More indications for lepton nonuniversality in $b \to s\ell^+\ell^-$. Phys. Lett. B **824**, 136838 (2022). https://doi.org/10.1016/j.physletb.2021.136838. arXiv:2104.10058
22. M. Algueró, B. Capdevila, S. Descotes-Genon, J. Matias, M. Novoa-Brunet, $b \to s\ell^+\ell^-$ global fits after R_{K_S} and $R_{K^{*+}}$. Eur. Phys. J. C **82**, 326 (2022). https://doi.org/10.1140/epjc/s10052-022-10231-1. arXiv:2104.08921
23. W. Altmannshofer, P. Stangl, New physics in rare B decays after Moriond 2021. Eur. Phys. J. C **81**, 952 (2021). https://doi.org/10.1140/epjc/s10052-021-09725-1. arXiv:2103.13370
24. M. Ciuchini, M. Fedele, E. Franco, A. Paul, L. Silvestrini, M. Valli, Lessons from the $B^{0,+} \to K^{*0,+}\mu^+\mu^-$ angular analyses. Phys. Rev. D **103**, 015030 (2021). https://doi.org/10.1103/PhysRevD.103.015030. arXiv:2011.01212
25. L.-S. Geng, B. Grinstein, S. Jäger, S.-Y. Li, J. Martin Camalich, R.-X. Shi, Implications of new evidence for lepton-universality violation in $b \to s\ell^+\ell^-$ decays. Phys. Rev. D **104**, 035029 (2021) https://doi.org/10.1103/PhysRevD.104.035029. arXiv:2103.12738
26. F. Mahmoudi, Theoretical review of rare B decays, in *20th Conference on Flavor Physics and CP Violation*, vol. 8 (2022). arXiv:2208.05755
27. ATLAS collaboration, Study of the rare decays of B_s^0 and B^0 mesons into muon pairs using data collected during 2015 and 2016 with the ATLAS detector. JHEP **04**, 098 (2019). https://doi.org/10.1007/JHEP04(2019)098. arXiv:1812.03017
28. CMS collaboration, Measurement of properties of $\mathrm{B_s^0} \to \mu^+\mu^-$ decays and search for $\mathrm{B}^0 \to \mu^+\mu^-$ with the CMS experiment. JHEP **04**, 188 (2020). https://doi.org/10.1007/JHEP04(2020)188. arXiv:1910.12127
29. LHCb collaboration, Measurement of the $B_s^0 \to \mu^+\mu^-$ decay properties and search for the $B^0 \to \mu^+\mu^-$ and $B_s^0 \to \mu^+\mu^-\gamma$ decays. Phys. Rev. D **105**, 012010 (2022). https://doi.org/10.1103/PhysRevD.105.012010. arXiv:2108.09283
30. LHCb collaboration, Analysis of neutral B-meson decays into two muons. Phys. Rev. Lett. **128**, 041801 (2022). https://doi.org/10.1103/PhysRevLett.128.041801. arXiv:2108.09284
31. CMS collaboration, Measurement of $B_s^0 \to \mu^+\mu^-$ decay properties and search for the $B^0 \to \mu\mu$ decay in proton-proton collisions at $\sqrt{s} = 13$ TeV. arXiv:CMS-PAS-BPH-21-006
32. T. Hurth, F. Mahmoudi, D. Martinez Santos, S. Neshatpour, Lepton nonuniversality in exclusive $b \to s\ell\ell$ decays. Phys. Rev. D **96**, 095034 (2017). https://doi.org/10.1103/PhysRevD.96.095034. arXiv:1705.06274
33. A. Arbey, T. Hurth, F. Mahmoudi, D.M. Santos, S. Neshatpour, Update on the $b \to s$ anomalies. Phys. Rev. D **100**, 015045 (2019). https://doi.org/10.1103/PhysRevD.100.015045. arXiv:1904.08399

34. F. Mahmoudi, SuperIso: a program for calculating the isospin asymmetry of $B \to K^*\gamma$ in the MSSM. Comput. Phys. Commun. **178**, 745 (2008). https://doi.org/10.1016/j.cpc.2007.12.006. arXiv:0710.2067
35. F. Mahmoudi, SuperIso v2.3: a program for calculating flavor physics observables in supersymmetry. Comput. Phys. Commun. **180**, 1579 (2009). https://doi.org/10.1016/j.cpc.2009.02.017. arXiv:0808.3144
36. F. Mahmoudi, SuperIso v3.0, flavor physics observables calculations: extension to NMSSM. Comput. Phys. Commun. **180**, 1718 (2009). https://doi.org/10.1016/j.cpc.2009.05.001
37. S. Neshatpour, F. Mahmoudi, Flavour physics with superIso. PoS **TOOLS2020**, 036 (2021). https://doi.org/10.22323/1.392.0036. arXiv:2105.03428
38. S. Neshatpour, F. Mahmoudi, Flavour physics phenomenology with superIso. PoS **CompTools2021**, 010 (2022). https://doi.org/10.22323/1.409.0010. arXiv:2207.04956
39. BELLE collaboration, Test of lepton flavor universality and search for lepton flavor violation in $B \to K\ell\ell$ decays. JHEP **03**, 105 (2021). https://doi.org/10.1007/JHEP03(2021)105. arXiv:1908.01848
40. T. Hurth, F. Mahmoudi, S. Neshatpour, Implications of the new LHCb angular analysis of $B \to K^*\mu^+\mu^-$?: hadronic effects or new physics? Phys. Rev. D **102**, 055001 (2020). https://doi.org/10.1103/PhysRevD.102.055001. arXiv:2006.04213
41. V.G. Chobanova, T. Hurth, F. Mahmoudi, D. Martinez Santos, S. Neshatpour, Large hadronic power corrections or new physics in the rare decay $B \to K^*\mu^+\mu^-$? JHEP **07**, 025 (2017). https://doi.org/10.1007/JHEP07(2017)025. arXiv:1702.02234
42. S. Neshatpour, V.G. Chobanova, T. Hurth, F. Mahmoudi, D. Martinez Santos, Direct comparison of global fits to the $B \to K^*\mu^+\mu^-$ data assuming hadronic corrections or new physics, in *52nd Rencontres de Moriond on QCD and High Energy Interactions* (2017), pp. 87–90. arXiv:1705.10730
43. CMS collaboration, Angular analysis of the decay $\mathrm{B}^+ \to \mathrm{K}^+\mu^+\mu^-$ in proton-proton collisions at $\sqrt{s} = 8$ TeV. Phys. Rev. D **98**, 112011 (2018). https://doi.org/10.1103/PhysRevD.98.112011. arXiv:1806.00636
44. LHCb collaboration, Strong constraints on the $b \to s\gamma$ photon polarisation from $B^0 \to K^{*0}e^+e^-$ decays. JHEP **12**, 081 (2020). https://doi.org/10.1007/JHEP12(2020)081. arXiv:2010.06011
45. T. Hurth, F. Mahmoudi, S. Neshatpour, On the anomalies in the latest LHCb data. Nucl. Phys. B **909**, 737 (2016). https://doi.org/10.1016/j.nuclphysb.2016.05.022. arXiv:1603.00865
46. T. Hurth, F. Mahmoudi, S. Neshatpour, Global fits to $b \to s\ell\ell$ data and signs for lepton non-universality. JHEP **12**, 053 (2014). https://doi.org/10.1007/JHEP12(2014)053. arXiv:1410.4545
47. A. Arbey, T. Hurth, F. Mahmoudi, S. Neshatpour, Hadronic and new physics contributions to $b \to s$ transitions. Phys. Rev. D **98**, 095027 (2018). https://doi.org/10.1103/PhysRevD.98.095027. arXiv:1806.02791

Chapter 3
Light Meson Spectroscopy and Gluonium Searches in η_c and $\Upsilon(1S)$ Decays at BaBar

Antimo Palano

Abstract We study the $\Upsilon(1S)$ radiative decays to $\gamma\pi^+\pi^-$ and γK^+K^- using data recorded with the BaBar detector operating at the SLAC PEP-II asymmetric-energy e^+e^- collider at center-of-mass energies at the $\Upsilon(2S)$ and $\Upsilon(3S)$ resonances. The $\Upsilon(1S)$ resonance is reconstructed from the decay $\Upsilon(nS) \to \pi^+\pi^-\Upsilon(1S), n = 2, 3$. We also study the processes $\gamma\gamma \to \eta_c \to \eta' K^+K^-$, $\eta'\pi^+\pi^-$, and $\eta\pi^+\pi^-$ using a data sample of 519 fb^{-1} recorded with the BaBar detector at center-of-mass energies at and near the $\Upsilon(nS)$ $(n = 2, 3, 4)$ resonances. A Dalitz plot analysis is performed of η_c decays to $\eta' K^+K^-$, $\eta'\pi^+\pi^-$, and $\eta\pi^+\pi^-$. We compare η_c decays to η and η' final states in association with scalar mesons as they relate to the identification of the scalar glueball.

3.1 Introduction

The existence of gluonium states is still an open issue for Quantum Chromodynamics (QCD). Lattice QCD calculations predict the lightest gluonium states to have quantum numbers $J^{PC} = 0^{++}$ and 2^{++} and to be in the mass region below 2.5 GeV/c^2 [1]. In particular, the $J^{PC} = 0^{++}$ glueball is predicted to have a mass around 1.7 GeV/c^2. The broad $f_0(500)$, $f_0(1370)$ [2], $f_0(1500)$ [3, 4], $f_0(1710)$ [5, 6] and possibly the $f_0(2100)$ [7] have been suggested as scalar glueball candidates. However, the identification of the scalar glueball is complicated by the possible mixing with standard $q\bar{q}$ states.

Radiative decays of heavy quarkonia, in which a photon replaces one of the three gluons from the strong decay of J/ψ or $\Upsilon(1S)$, can probe color-singlet two-gluon systems that produce gluonic resonances. J/ψ decays have been extensively

On behalf of the BaBar Collaboration.

A. Palano (✉)
INFN, Sezione di Bari, Bari, Italy
e-mail: antimo.palano@ba.infn.it

G. Ricciardi et al. (eds.), *8th Workshop on Theory, Phenomenology and Experiments in Flavour Physics*, Springer Proceedings in Physics 292,
https://doi.org/10.1007/978-3-031-30459-0_3

studied [8, 9]. In the first BaBar analysis [10] summarized in the present review, we study $\Upsilon(1S)$ decays, taking into account that the experimental observation of radiative $\Upsilon(1S)$ decays is challenging because their rate is suppressed by a factor of $\approx$0.025 compared to J/ψ radiative decays, which are of order 10^{-3} [11].

Decays of the η_c, the lightest pseudoscalar $c\bar{c}$ state, provide a window on light meson states. In the second analysis [12] summarized in the present review, we consider the three-body η_c decays to $\eta' K^+K^-$, $\eta'\pi^+\pi^-$, and $\eta\pi^+\pi^-$, using two-photon interactions, $e^+e^- \rightarrow e^+e^-\gamma^*\gamma^* \rightarrow e^+e^-\eta_c$. If both of the virtual photons are quasi-real, the allowed J^{PC} values of any produced resonances are $0^{\pm+}$, $2^{\pm+}$, $4^{\pm+}$... [13]. The possible presence of a gluonic component of the η' meson, due to the so-called gluon anomaly, has been discussed in recent years [14, 15]. A comparison of the η and η' content of η_c decays might yield information on the possible gluonic content of resonances decaying to $\pi^+\pi^-$ or K^+K^-.

3.2 Study of $\Upsilon(1S)$ Radiative Decays to $\gamma\pi^+\pi^-$ and γK^+K^-

3.2.1 *Events Reconstruction*

We reconstruct the decay chains

$$\Upsilon(2S)/\Upsilon(3S) \rightarrow (\pi_s^+\pi_s^-)\Upsilon(1S) \rightarrow (\pi_s^+\pi_s^-)(\gamma\pi^+\pi^-) \tag{3.1}$$

and

$$\Upsilon(2S)/\Upsilon(3S) \rightarrow (\pi_s^+\pi_s^-)\Upsilon(1S) \rightarrow (\pi_s^+\pi_s^-)(\gamma K^+K^-), \tag{3.2}$$

where we label with the subscript s the slow pions from the direct $\Upsilon(2S)$ and $\Upsilon(3S)$ decays.

Events with balanced momentum are required to satisfy energy balance requirements. For each combination of $\pi_s^+\pi_s^-$ candidates, we first require both particles to be identified loosely as pions and compute the recoiling mass

$$M^2_{\text{rec}}(\pi_s^+\pi_s^-) = |p_{e^+} + p_{e^-} - p_{\pi_s^+} - p_{\pi_s^-}|^2, \tag{3.3}$$

where p is the particle four-momentum. The distribution of $M^2_{\text{rec}}(\pi_s^+\pi_s^-)$ is expected to peak at the squared $\Upsilon(1S)$ mass for signal events. Figure 3.1 shows the combinatorial recoiling mass $M_{\text{rec}}(\pi_s^+\pi_s^-)$ for $\Upsilon(2S)$ and $\Upsilon(3S)$ data, where narrow peaks at the $\Upsilon(1S)$ mass can be observed.

We select signal event candidates by requiring

$$|M_{\text{rec}}(\pi_s^+\pi_s^-) - m(\Upsilon(1S))_f| < 2.5\sigma, \tag{3.4}$$

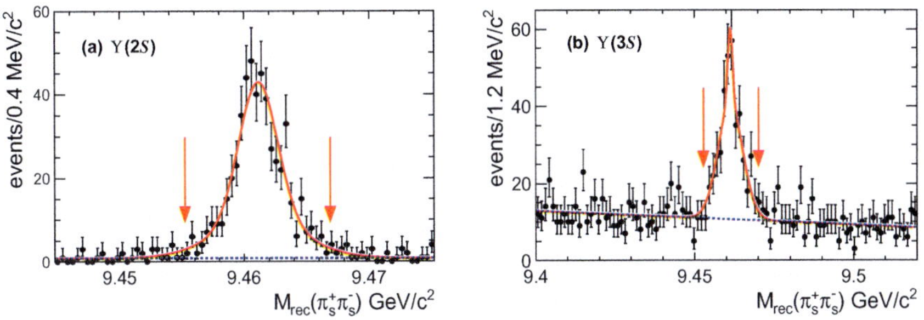

Fig. 3.1 Combinatorial recoiling mass M_{rec} to $\pi_s^+\pi_s^-$ candidates for **a** $\Upsilon(2S)$ and **b** $\Upsilon(3S)$ data. The arrows indicate the regions used to select the $\Upsilon(1S)$ signal

where $m(\Upsilon(1S))_f$ indicates the fitted $\Upsilon(1S)$ mass value and $\sigma = 2.3$ MeV/c^2 and σ=3.5 MeV/c^2 for $\Upsilon(2S)$ and $\Upsilon(3S)$ data, respectively. To reconstruct $\Upsilon(1S) \to \gamma\pi^+\pi^-$ or $\Upsilon(1S) \to \gamma K^+K^-$ decays, we require a loose identification of both pions or kaons and isolate the two $\Upsilon(1S)$ decay modes by requiring

$$9.1 \text{ GeV}/c^2 < m(\gamma h^+h^-) < 9.6 \text{ GeV}/c^2, \tag{3.5}$$

where $h = \pi, K$.

3.2.2 *Study of the $\pi^+\pi^-$ and K^+K^- Mass Spectra*

The $\pi^+\pi^-$ mass spectrum, for $m(\pi^+\pi^-) < 3.0$ GeV/c^2 and summed over the $\Upsilon(2S)$ and $\Upsilon(3S)$ datasets with 507 and 277 events, respectively, is shown in Fig. 3.2(Left). The spectrum shows $I = 0$, $J^P = \text{even}^{++}$ resonance production, with low backgrounds above 1 GeV/c^2. We observe a rapid drop around 1 GeV/c^2 characteristic of the presence of the $f_0(980)$, and a strong $f_2(1270)$ signal. The data also suggest the presence of additional weaker resonant contributions.

The K^+K^- mass spectrum, summed over the $\Upsilon(2S)$ and $\Upsilon(3S)$ datasets with 164 and 63 events, respectively, is shown in Fig. 3.2(Right) and also shows resonant production, with low background. Signals at the positions of $f_2'(1525)/f_0(1500)$ and $f_0(1710)$ can be observed, with further unresolved structure at higher mass.

We make use of a phenomenological model to extract the different $\Upsilon(1S) \to \gamma R$ branching fractions, where R is an intermediate resonance. We perform a simultaneous binned fit to the $\pi^+\pi^-$ mass spectra from the $\Upsilon(2S)$ and $\Upsilon(3S)$ datasets. We describe the low-mass region (around the $f_0(500)$) using a relativistic S-wave Breit–Wigner lineshape having free parameters. We describe the $f_0(980)$ using the Flatté [16] formalism with parameters fixed to the values from [17]. The $f_2(1270)$ and $f_0(1710)$ resonances are represented by relativistic Breit–Wigner functions with parameters fixed to PDG values [18]. In the high $\pi^+\pi^-$ mass region we include a

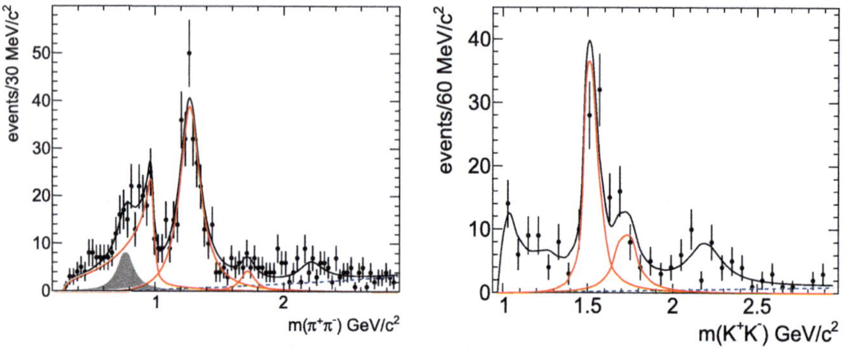

Fig. 3.2 (Left) $\pi^+\pi^-$ mass distribution from $\Upsilon(1S) \to \pi^+\pi^-\gamma$ for the combined $\Upsilon(2S)$ and $\Upsilon(3S)$ datasets. The full (red) curves indicate the S-wave, $f_2(1270)$, and $f_0(1710)$ contributions. The shaded (gray) area represents the estimated $\rho(770)^0$ background. (Right) K^+K^- mass distribution from $\Upsilon(1S) \to K^+K^-\gamma$ for the combined $\Upsilon(2S)$ and $\Upsilon(3S)$ datasets. The (red) curves show the contributions from $f_2'(1525)/f_0(1500)$ and $f_0(1710)$. Dashed (blue) lines indicate the background contributions

single resonance $f_0(2100)$ having a width fixed to the PDG value (224 ± 22) and unconstrained mass. For the $\Upsilon(3S)$ data we also include $\rho(770)^0$ background with parameters fixed to the PDG values. The fit is shown in Fig. 3.2. It has 16 free parameters and $\chi^2 = 182$ for ndf $= 152$, corresponding to a p-value of 5%. We note the observation of a significant S-wave in $\Upsilon(1S)$ radiative decays. This observation was not possible in the study of J/ψ radiative decay to $\pi^+\pi^-$ because of the presence of a strong, irreducible background from $J/\psi \to \pi^+\pi^-\pi^0$ [19]. No evidence is found for a $\Upsilon(1S) \to \pi^+\pi^-\pi^0$ decay in present data. We obtain the following $f_0(500)$ parameters:

$$m(f_0(500)) = 0.856 \pm 0.086 \text{ GeV}/c^2, \; \Gamma(f_0(500)) = 1.279 \pm 0.324 \text{ GeV}, \quad (3.6)$$

and $\phi = 2.41 \pm 0.43$ rad. The fraction of S-wave events associated with the $f_0(500)$ is $(27.7 \pm 3.1)\%$.

We perform a binned fit to the combined K^+K^- mass spectrum using the following model. The $f_0(980)$ is parameterized according to the Flatté formalism. The $f_2(1270)$, $f_2'(1525)$, $f_0(1500)$, and $f_0(1710)$ resonances are represented by relativistic Breit–Wigner functions with parameters fixed to PDG values. We include an $f_0(2200)$ contribution having parameters fixed to the PDG values. The fit shown in Fig. 3.2(Right). It has six free parameters and $\chi^2 = 35$ for ndf $= 29$, corresponding to a p-value of 20%. The resonances yields and significances are given in Table 3.1. Systematic uncertainties are dominated by the PDG uncertainties on resonances parameters.

The efficiency distributions as functions of mass, for the $\Upsilon(2S)/\Upsilon(3S)$ data and for the $\pi^+\pi^-\gamma$ and $K^+K^-\gamma$ final states, are found to have an almost uniform behavior for all the final states. We define the helicity angle θ_H as the angle formed by the

Table 3.1 Resonances yields and statistical significances from the fits to the $\pi^+\pi^-$ and K^+K^- mass spectra for the $\Upsilon(2S)$ and $\Upsilon(3S)$ datasets. The symbol $f_J(1500)$ indicates the signal in the 1500 MeV/c^2 mass region

Resonances ($\pi^+\pi^-$)	Yield $\Upsilon(2S)$	Yield $\Upsilon(3S)$	Significance
S-wave	$133 \pm 16 \pm 13$	87 ± 13	12.8σ
$f_2(1270)$	$255 \pm 19 \pm 8$	$77 \pm 7 \pm 4$	15.9σ
$f_0(1710)$	$24 \pm 8 \pm 6$	$6 \pm 8 \pm 3$	2.5σ
Resonances (K^+K^-)	Yield $\Upsilon(2S) + \Upsilon(3S)$		Significance
$f_0(980)$	47 ± 9		5.6σ
$f_J(1500)$	$77 \pm 10 \pm 10$		8.9σ
$f_0(1710)$	$36 \pm 9 \pm 6$		4.7σ

h^+, in the h^+h^- rest frame, and the γ in the $h^+h^-\gamma$ rest frame. We also define θ_γ as the angle formed by the radiative photon in the $h^+h^-\gamma$ rest frame with respect to the $\Upsilon(1S)$ direction in the $\Upsilon(2S)/\Upsilon(3S)$ rest frame. We label with $\epsilon(m, \cos\theta_H)$ the efficiency computed as a function of the h^+h^- effective mass and the helicity angle $\cos\theta_H$. We label with $\epsilon(\cos\theta_H, \cos\theta_\gamma)$ the efficiency computed, for each resonance mass window, as a function of $\cos\theta_H$ and $\cos\theta_\gamma$. This is used to obtain the efficiency-corrected angular distributions and branching fractions for the different resonances. To obtain the efficiency correction weight w_R for the resonance R we divide each event by the efficiency $\epsilon(\cos\theta_H, \cos\theta_\gamma)$

$$w_R = \frac{\sum_{i=1}^{N_R} 1/\epsilon_i(\cos\theta_H, \cos\theta_\gamma)}{N_R}, \tag{3.7}$$

where N_R is the number of events in the resonance mass range.

3.2.3 Angular Analysis

To obtain information on the angular momentum structure of the $\pi^+\pi^-$ and K^+K^- systems in $\Upsilon(1S) \to \gamma h^+h^-$ we study the dependence of the $m(h^+h^-)$ mass on the helicity angle θ_H. A better way to observe angular effects is to plot the $\pi^+\pi^-$ mass spectrum weighted by the Legendre polynomial moments, corrected for efficiency and shown in Fig. 3.3. In a simplified environment, the moments are related to the spin 0 (S) and spin 2 (D) amplitudes by the equations

$$\begin{aligned} \sqrt{4\pi}\langle Y_0^0 \rangle &= S^2 + D^2, \\ \sqrt{4\pi}\langle Y_2^0 \rangle &= 2SD\cos\phi_{SD} + 0.639D^2, \\ \sqrt{4\pi}\langle Y_4^0 \rangle &= 0.857D^2, \end{aligned} \tag{3.8}$$

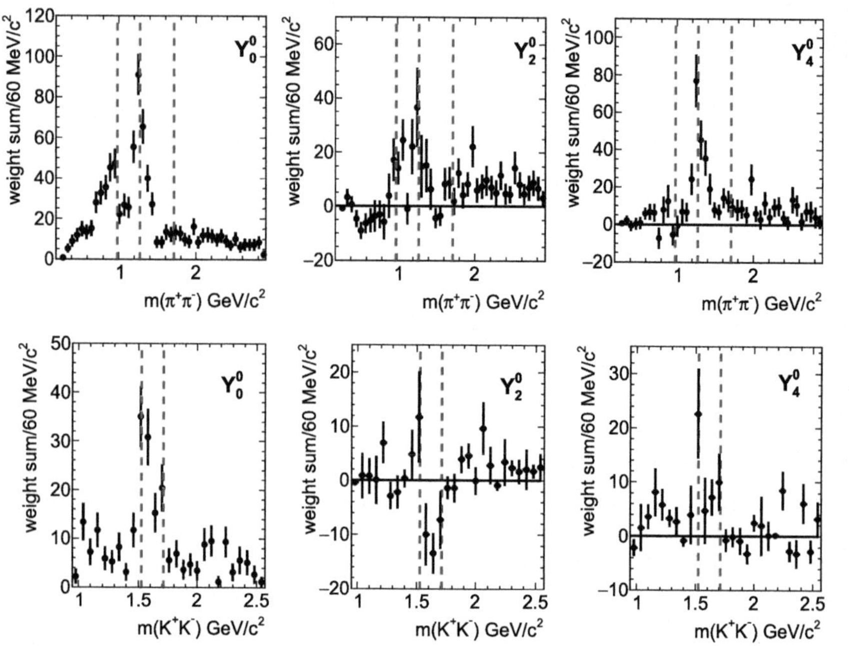

Fig. 3.3 The distributions of the most relevant unnormalized Y_L^0 moments for $\Upsilon(1S) \to \gamma\pi^+\pi^-$ (Top) and $\Upsilon(1S) \to \gamma K^+K^-$ (Bottom) as functions of mass. The lines indicate the positions of $f_0(980)$, $f_2(1270)$, and $f_0(1710)$ for $\pi^+\pi^-$ and $f_2'(1525)$ and $f_0(1710)$ for K^+K^-

where ϕ_{SD} is the relative phase. Therefore we expect to observe spin 2 resonances in $\langle Y_4^0\rangle$ and S/D interference in $\langle Y_2^0\rangle$. The results are shown in Fig. 3.3(Top). We clearly observe the $f_2(1270)$ resonance in $\langle Y_4^0\rangle$ and a sharp drop in $\langle Y_2^0\rangle$ at the $f_2(1270)$ mass, indicating the interference effect. The distribution of $\langle Y_0^0\rangle$ is just the scaled $\pi^+\pi^-$ mass distribution, corrected for efficiency. Similarly, we plot in Fig. 3.3(Bottom) the K^+K^- mass spectrum weighted by the Legendre polynomial moments, corrected for efficiency. We observe signals of the $f_2'(1525)$ and $f_0(1710)$ in $\langle Y_4^0\rangle$ and activity due to S/D interference effects in the $\langle Y_2^0\rangle$ moment.

Resonance angular distributions in radiative $\Upsilon(1S)$ decays from $\Upsilon(2S)/\Upsilon(3S)$ decays are rather complex (see [10] for details). Here we only perform a simplified Partial Wave Analysis (PWA) solving directly the system of (3.8). Figure 3.4 shows the resulting S-wave and D-wave contributions to the $\pi^+\pi^-$ and K^+K^- mass spectra, respectively. Due to the presence of background in the threshold region, the $\pi^+\pi^-$ analysis is performed only on the $\Upsilon(2S)$ data.

We note that in the case of the $\pi^+\pi^-$ mass spectrum we obtain a good separation between S and D-waves, with the presence of an $f_0(980)$ resonance on top of a broad $f_0(500)$ resonance in the S-wave and a clean $f_2(1270)$ in the D-wave distribution. Integrating the S-wave amplitude from threshold up to a mass of 1.5 GeV/c^2, we obtain an integrated, efficiency corrected yield $N(S\text{–wave}) = 629 \pm 128$.

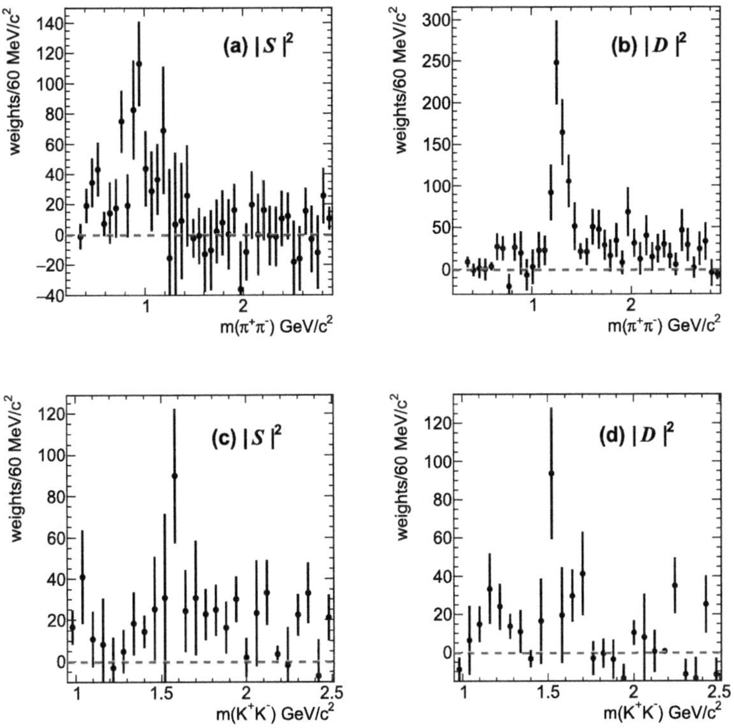

Fig. 3.4 **a–c** S and **b–d** D-wave contributions from the simple PWA of the $\pi^+\pi^-$ mass spectrum for the $\Upsilon(2S)$ data (Left) and of the K^+K^- mass spectrum for the combined $\Upsilon(2S)/\Upsilon(3S)$ data (Right)

In the case of the K^+K^- PWA the structure peaking around 1500 MeV/c^2 appears in both S and D-waves suggesting the presence of $f_0(1500)$ and $f_2'(1525)$. In the $f_0(1710)$ mass region there is not enough data to discriminate between the two different spin assignments. This pattern is similar to that observed in the Dalitz plot analysis of charmless $B \to 3K$ decays [20]. Integrating the S and D-wave contributions in the $f_2'(1525)/f_0(1500)$ mass region, we obtain a fraction of S-wave contribution $f_S(K^+K^-) = 0.53 \pm 0.10$.

3.2.4 Measurement of Branching Fractions

We determine the branching fraction $\mathcal{B}(R)$ for the decay of $\Upsilon(1S)$ to photon and resonance R using the expression

$$\mathcal{B}(R) = \frac{N_R(\Upsilon(nS) \to \pi_s^+\pi_s^-\Upsilon(1S)(\to R\gamma))}{N(\Upsilon(nS) \to \pi_s^+\pi_s^-\Upsilon(1S)(\to \mu^+\mu^-))} \times \mathcal{B}(\Upsilon(1S) \to \mu^+\mu^-), \quad (3.9)$$

where N_R indicates the efficiency-corrected yield for the given resonance. To reduce systematic uncertainties, we first compute the relative branching fraction to the reference channel $\Upsilon(nS) \to \pi^+\pi^-\Upsilon(1S)(\to \mu^+\mu^-)$, which has the same number of charged particles as the final states under study. We then multiply the relative branching fraction by the well-measured branching fraction $\mathcal{B}(\Upsilon(1S) \to \mu^+\mu^-) = 2.48 \pm 0.05\%$ [18].

We determine the reference channel corrected yield using the method of "B-counting", also used to obtain the number of produced $\Upsilon(2S)$ and $\Upsilon(3S)$ [21]. Taking into account the known branching fractions of $\Upsilon(2S)/\Upsilon(3S) \to \pi_s^+\pi_s^-\Upsilon(1S)$ we obtain

$$N(\Upsilon(2S) \to \pi_s^+\pi_s^-\Upsilon(1S)(\to \mu^+\mu^-)) = (4.35 \pm 0.12_{\text{sys}}) \times 10^5 \tag{3.10}$$

and

$$N(\Upsilon(3S) \to \pi_s^+\pi_s^-\Upsilon(1S)(\to \mu^+\mu^-)) = (1.32 \pm 0.04_{\text{sys}}) \times 10^5 \tag{3.11}$$

events. Table 3.2 gives the measured branching fractions. In all cases we correct the efficiency corrected yields for isospin and for PDG measured branching fractions [18].

We report the first observation of $f_0(1710)$ in $\Upsilon(1S)$ radiative decay with a significance of 5.7σ, combining $\pi^+\pi^-$ and K^+K^- data. To determine the branching ratio of the $f_0(1710)$ decays to $\pi\pi$ and $K\bar{K}$, we remove all the systematic uncertainties related to the reference channels and of the γ reconstruction and obtain

$$\frac{\mathcal{B}(f_0(1710) \to \pi\pi)}{\mathcal{B}(f_0(1710) \to K\bar{K})} = 0.64 \pm 0.27_{\text{stat}} \pm 0.18_{\text{sys}}, \tag{3.12}$$

in agreement with the world average value of $0.41^{+0.11}_{-0.17}$ [18].

Table 3.2 Measured $\Upsilon(1S) \to \gamma R$ branching fractions

Resonance	$\mathcal{B}(10^{-5})$
$\pi\pi$ S-wave	$4.63 \pm 0.56 \pm 0.48$
$f_2(1270)$	$10.15 \pm 0.59\ ^{+0.54}_{-0.43}$
$f_0(1710) \to \pi\pi$	$0.79 \pm 0.26 \pm 0.17$
$f_J(1500) \to K\bar{K}$	$3.97 \pm 0.52 \pm 0.55$
$f_2'(1525)$	$2.13 \pm 0.28 \pm 0.72$
$f_0(1500) \to K\bar{K}$	$2.08 \pm 0.27 \pm 0.65$
$f_0(1710) \to K\bar{K}$	$2.02 \pm 0.51 \pm 0.35$

3.3 Dalitz Plot Analysis of η_c Three-Body Decays

The results presented here are based on the full data set collected with the BaBar detector using an integrated luminosity of 519 fb^{-1}recorded at center-of-mass energies at and near the $\Upsilon(nS)$ $(n = 2, 3, 4)$ resonances. In the present analysis, we consider the three-body η_c decays to $\eta' K^+K^-$, $\eta'\pi^+\pi^-$, and $\eta\pi^+\pi^-$, using two-photon interactions, $e^+e^- \to e^+e^-\gamma^*\gamma^* \to e^+e^-\eta_c$, where γ^* indicate the intermediate quasi-real virtual photons.

3.3.1 Study of $\gamma\gamma \to \eta' h^+h^-$ and $\gamma\gamma \to \eta\pi^+\pi^-$

We first study the reactions

$$\gamma\gamma \to \eta' h^+h^-, \tag{3.13}$$

where h^+h^- indicates a $\pi^+\pi^-$ or K^+K^- system. The η' is reconstructed in the two decay modes $\eta' \to \rho^0\gamma$, $\rho^0 \to \pi^+\pi^-$, and $\eta' \to \eta\pi^+\pi^-$, $\eta \to \gamma\gamma$. We define p_T as the magnitude of the transverse momentum of the $\eta' h^+h^-$ system, in the e^+e^- rest frame, with respect to the beam axis. Well reconstructed two-photon events with quasi-real photons are expected to have low values of p_T. For the selection of the $\eta'\pi^+\pi^-$ final state, we require all four charged tracks to be positively identified as pions. For the $\eta' K^+K^-$ final state, we require the two charged tracks assigned to the η' decay to be positively identified as pions and the other two to be positively identified as kaons. We require $p_T < 0.05$ GeV/c and $p_T < 0.15$ GeV/c, for the $\eta' \to \rho^0\gamma$ and $\eta' \to \eta\pi^+\pi^-$, respectively. We discriminate against Initial State Radiation (ISR) events $e^+e^- \to \gamma_{ISR}\eta'(\eta)h^+h^-$, by requiring the recoil mass $M_{\text{rec}} \equiv (p_{e^+e^-} - p_{\text{rec}})^2 > 2$ GeV$^2/c^4$, where $p_{e^+e^-}$ is the four-momentum of the initial state e^+e^- and p_{rec} is the reconstructed four-momentum of the candidate $\eta'(\eta)h^+h^-$ system.

The $\eta'\pi^+\pi^-$ and $\eta' K^+K^-$ mass spectra, summed over the $\eta' \to \rho^0\gamma$ and $\eta' \to \eta\pi^+\pi^-$ decay modes are shown in Fig. 3.5, where prominent η_c signals can be observed. In particular, Fig. 3.5(Right) reports the first observation of the decay $\eta_c \to \eta' K^+K^-$.

We also study the reaction

$$\gamma\gamma \to \eta\pi^+\pi^-, \tag{3.14}$$

where $\eta \to \gamma\gamma$ and $\eta \to \pi^+\pi^-\pi^0$. In this case the two-photon reaction is selected by requiring $p_T < 0.1$ GeV/c for both η decay modes. The corresponding $\eta\pi^+\pi^-$ mass spectra are shown in Fig. 3.6, where prominent η_c signals can be observed.

To compute the reconstruction and selection efficiency, MC signal events are generated using a detailed detector simulation in which the η_c mesons decay uniformly in phase space. These simulated events are reconstructed and analyzed in the same

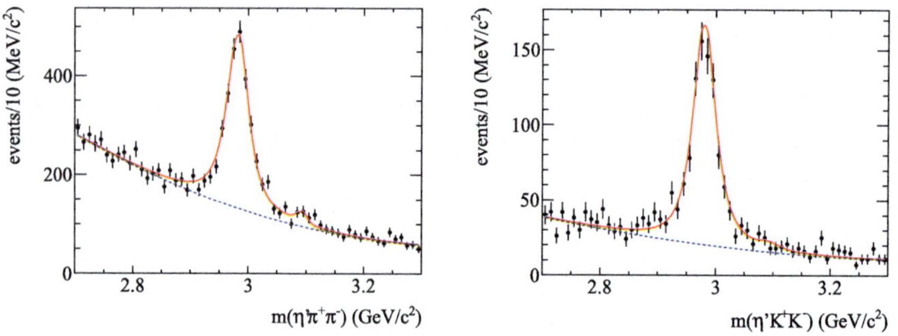

Fig. 3.5 Invariant-mass distributions of selected (Left) $\eta'\pi^+\pi^-$ and (Right) $\eta' K^+K^-$ candidates summed over the $\eta' \to \rho^0\gamma$ and $\eta' \to \eta\pi^+\pi^-$ decay modes

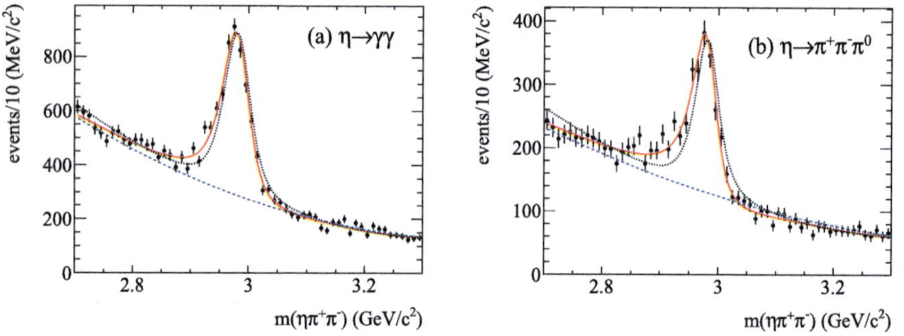

Fig. 3.6 Invariant-mass spectra for selected $\eta\pi^+\pi^-$ candidate events with **a** $\eta \to \gamma\gamma$ and **b** $\eta \to \pi^+\pi^-\pi^0$. The solid (red) lines represent the fits including interference described in the text. The dashed (blue) line represents the fitted non-resonant components. The dotted lines represent the fits without interference

manner as data. We define the helicity angle θ_H as the angle formed by the h^+ (where $h = \pi, K$), in the h^+h^- rest frame, and the η' (η) direction in the $h^+h^-\eta'$ ($h^+h^-\eta$) rest frame. To smoothen statistical fluctuations, the efficiency maps are parameterized using Legendre polynomials up to $L = 12$ as functions of $\cos\theta_H$ in intervals of $m(h^+h^-)$ and then interpolated linearly between adjacent mass intervals.

3.3.2 Yields and Branching Fractions

We fit the invariant-mass distributions to obtain the numbers of selected η_c events, $N_{\eta' K^+K^-}$, $N_{\eta'\pi^+\pi^-}$, and $N_{\eta\pi^+\pi^-}$, for each η' or η decay mode. We then use the $\eta' K^+K^-$ and $\eta'\pi^+\pi^-$ yields to compute the ratio of branching fractions for η_c to the $\eta' K^+K^-$ and $\eta'\pi^+\pi^-$ final states. This ratio is computed as

$$\mathcal{R} = \frac{\mathcal{B}(\eta_c \to \eta' K^+ K^-)}{\mathcal{B}(\eta_c \to \eta' \pi^+ \pi^-)} = \frac{N_{\eta' K^+ K^-}}{N_{\eta' \pi^+ \pi^-}} \frac{\epsilon_{\eta' \pi^+ \pi^-}}{\epsilon_{\eta' K^+ K^-}} \tag{3.15}$$

for each η' decay mode, where $\epsilon_{\eta' K^+ K^-}$ and $\epsilon_{\eta' \pi^+ \pi^-}$ are the corresponding efficiencies. We determine $N_{K^+ K^- \eta'}$ and $N_{\pi^+ \pi^- \eta'}$ from η_c decays by performing binned χ^2 fits to the $\eta' K^+ K^-$ and $\eta' \pi^+ \pi^-$ invariant-mass spectra, in the 2.7–3.3 GeV/c^2 mass region, separately for the two η' decay modes. In these fits, the η_c signal contribution is described by a simple Breit–Wigner (BW) function convolved with a fixed resolution function (described by the sum of a Gaussian and Crystal Ball functions), with η_c parameters fixed to PDG values [22]. An additional BW function is used to describe the residual background from ISR J/ψ events, and the remaining background is parameterized by a 2nd order polynomial. The fitted $\eta' h^+ h^-$ invariant-mass spectra are shown in Fig. 3.5 summed over the two η' decay modes.

We estimate $\epsilon_{\eta' K^+ K^-}$ and $\epsilon_{\eta' \pi^+ \pi^-}$ for the η_c signals using the 2-D efficiency functions described above. Each event is first weighted by $1/\epsilon(m, \cos\theta_H)$. Since the backgrounds below the η_c signals have different distributions in the Dalitz plot, we perform a sideband subtraction by assigning an additional weight of $+1$ to events in the η_c signal region, defined as the (2.93–3.03) GeV/c^2 mass region, and a weight -1 to events in the sideband regions, (2.77–2.87) GeV/c^2 and (3.09–3.19) GeV/c^2. The two evaluations of the branching fractions, for the two η' decay mode, are in good agreement and give an average value of

$$\frac{\mathcal{B}(\eta_c \to \eta' K^+ K^-)}{\mathcal{B}(\eta_c \to \eta' \pi^+ \pi^-)} = 0.644 \pm 0.039_{\text{stat}} \pm 0.032_{\text{sys}}. \tag{3.16}$$

For the decay $\eta_c \to \eta \pi^+ \pi^-$ the fits without interference do not describe the data well for either η decay mode. Leaving free the η_c parameters, the fits return masses shifted down by $\approx$10 MeV/c^2 with respect to PDG averages.

We test the possibility of interference effects of the η_c with each non-resonant two-photon process [23], modifying the fitting function by defining

$$f(m) = |A_{\text{nres}}|^2 + |A_{\eta_c}|^2 + c \cdot 2Re(A_{\text{nres}} A^*_{\eta_c}), \tag{3.17}$$

where A_{nres} is the non-resonant amplitude with $|A_{\text{nres}}|^2$ described by a 2nd order polynomial; the coherence factor c is the fraction of the non-resonant events that are true two-photon production of the same final state; the resonant contribution is described by $A_{\eta_c} = \alpha \cdot BW(m) \cdot \exp(i\phi)$, where $BW(m)$ is a simple Breit–Wigner with parameters fixed to PDG values; and α, ϕ, and c are free parameters. The fitted invariant-mass spectra are shown in Fig. 3.6, where reasonable descriptions of the data are evident. As a comparison we also show the fit the two mass spectra with no interference and fixed η_c parameters and obtain the dotted lines distributions shown in Fig. 3.6.

3.3.3 *Dalitz Plot Analysis*

We perform Dalitz plot analyses of the $\eta'\pi^+\pi^-$, $\eta' K^+K^-$, and $\eta\pi^+\pi^-$ systems in the η_c mass region using unbinned maximum likelihood fits. Amplitudes are parameterized as described in [24]. They include a relativistic Breit–Wigner function having a variable width modulated by the Blatt–Weisskopf [25] spin form factors and the relevant spin-angular information. We first fit the two η_c sidebands separately, using an incoherent sum of amplitudes. To model the background composition in the η_c signal region, we take a weighted average of the two fitted fractional contributions, and normalize using the results from the fit to the η_c signal region.

3.3.3.1 Dalitz Plot Analysis of $\eta_c \to \eta' K^+K^-$

Figure 3.7a shows the Dalitz plot for the selected $\eta_c \to \eta' K^+K^-$ candidates in the data, for the two η' decay modes combined (930 events).

We observe that this η_c decay mode is dominated by a diagonal band on the low mass side of the Dalitz plot. The $m(K^+K^-)$ spectrum shows a large structure in the region of the $f_0(1710)$ resonance. The combined $m(\eta' K^\pm)$ invariant-mass spectrum shows a structure at threshold due to the $K_0^*(1430)$ accompanied by weaker resonant structures. The $K_0^*(1430)$ is a relatively broad resonance decaying to $K\pi$, $K\eta$, and $K\eta'$. The measured $K\eta$ relative branching fraction is $\frac{\mathcal{B}(K_0^*(1430)\to K\eta)}{\mathcal{B}(K_0^*(1430)\to K\pi)} = 0.092 \pm 0.025^{+0.010}_{-0.025}$ [26], while the $K\eta'$ has only been observed in [27]. To describe the $K_0^*(1430)$ lineshape in the $K\eta'$ projection, we model it using a simplified coupled-channel Breit–Wigner function, which ignores the small $K\eta$ contribution. We parameterize the $K_0^*(1430)$ signal as

$$BW(m) = \frac{1}{m_0^2 - m^2 - i(\rho_1(m)g^2_{K\pi} + \rho_2(m)g^2_{K\eta'})}, \tag{3.18}$$

where m_0 is the resonance mass, $g_{K\pi}$ and $g_{K\eta'}$ are the couplings to the $K\pi$ and $K\eta'$ final states, and $\rho_j(m) = 2P/m$ are the respective Lorentz-invariant phase-space factors, with P the decay particle momentum in the $K_0^*(1430)$ rest frame. The values of m_0 and the g_{Kj} couplings cannot be derived from the $K\eta'$ system only, and therefore we make use of the $K\pi$ S-wave measurement from BaBar [28]. We average the reported quasi model-independent (QMI) measurements of the $K\pi$ S-wave from $\eta_c \to K_S^0 K\pi$ and $\eta_c \to K^+K^-\pi^0$ decays, and obtain the modulus squared of the amplitude and the phase shown in Fig. 3.8.

We perform a simultaneous binned χ^2 fit to the $K\pi$ S-wave amplitude and phase from threshold up to 1.72 GeV/c^2. We model the $K\pi$ S-wave in this region as S-wave$(m) = B(m) + c \cdot BW_{K\pi}(m)e^{i\phi}$, where $BW_{K\pi}(m)$ is given by (3.18), $B(m)$ is an empirical background term, parameterized as $B(m) = \rho_1(m)e^{-\alpha m}$, and c, ϕ, and α are free parameters. The results of the fit are shown in Fig. 3.8 as the solid (red) lines.

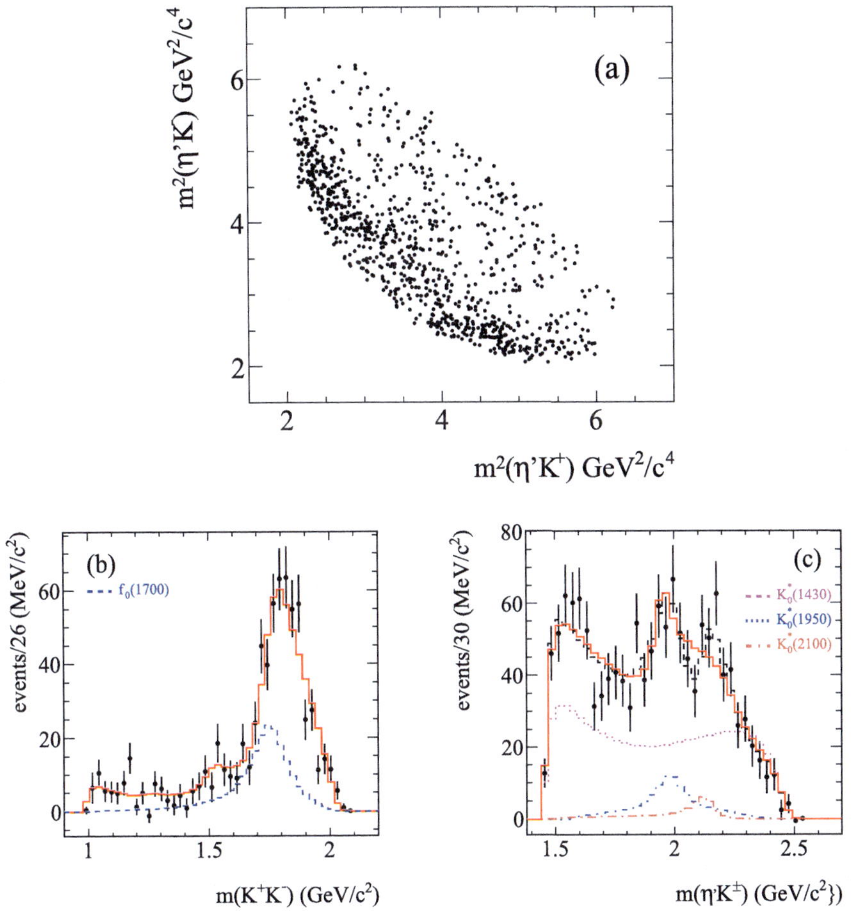

Fig. 3.7 **a** Dalitz plot for selected $\eta_c \to \eta' K^+ K^-$ candidates in the η_c signal region, summed over the two η' decay modes. Linear-scale mass projections **b** $m(K^+K^-)$ and **c** $m(\eta' K^\pm)$, after subtraction of the background. The solid (red) histograms represent the results of the fit described in the text. The (black) dashed line in (**c**) shows the solution which include the presence of $K_0^*(2130)$. The other histograms display the contributions from each of the listed components

We perform a Dalitz plot analysis of the $\eta_c \to \eta' K^+ K^-$ decay channel by using the $\eta' f_0(1710)$ intermediate state as the reference amplitude. The projections of the fit result are shown in Fig. 3.7b, c, along with the largest signal components. We measure the $f_0(1710)$ parameters, listed in Table 3.3. We also measure the parameters of the $K_0^*(1950)$ (see Table 3.3) for which there is only one previous measurement from the LASS collaboration [29]. For the $K_0^*(1430)$ resonance we make use of an iterative procedure, combining the results of the present Dalitz plot analysis with previous measurements (see [12] for details). The results from the Dalitz analysis are given in Table 3.4.

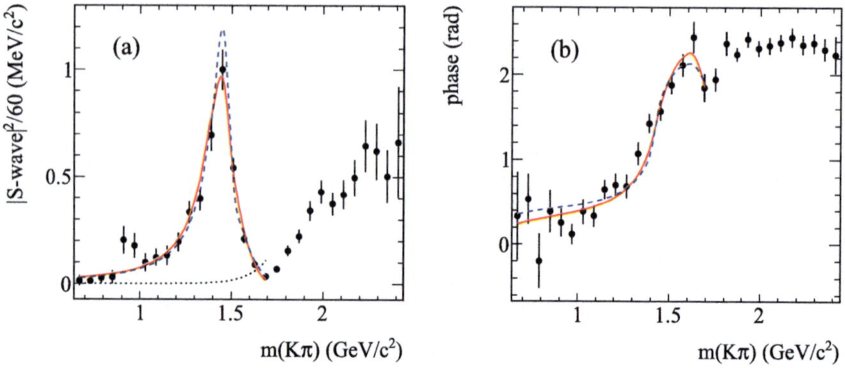

Fig. 3.8 The **a** squared modulus and **b** phase of the $K\pi$ S-wave averaged over the $\eta_c \rightarrow K^0_S K\pi$ and $\eta_c \rightarrow K^+K^-\pi^0$ from the BaBar [28] QMI analysis. The full (red) lines represent the result from the fit with free $g^2_{K\eta'}$ and $g^2_{K\pi}$ parameters. The dashed (blue) lines represent the result from the fit with a fixed $g^2_{K\eta'}/g^2_{K\pi}$ ratio [12]. The dotted (black) line in (**a**) represents the empirical background contribution

Table 3.3 Resonance parameters from the Dalitz plot analyses of $\eta_c \rightarrow \eta' K^+K^-$, $\eta_c \rightarrow \eta'\pi^+\pi^-$, and $\eta_c \rightarrow \eta\pi^+\pi^-$. Significances are computed using the Wilks theorem [30] and do not include systematic uncertainties

Resonance	Mass (MeV/c^2)	$g^2_{K\pi}$ (GeV^2/c^4)	$g^2_{K\eta'}$ (GeV^2/c^4)
		$\eta_c \rightarrow \eta' K^+K^-$	
$K^*_0(1430)$			
$\eta_c \rightarrow K\bar{K}\pi$	1447 ± 8	0.414 ± 0.026	0.197 ± 0.105
Fixed $\frac{g^2_{\eta' K}}{g^2_{\pi K}}$	1453 ± 22	0.462 ± 0.036	
Resonance	Mass (MeV/c^2)	Γ (MeV)	Significance ($n\sigma$)
$f_0(1710)$	$1757 \pm 24 \pm 9$	$175 \pm 23 \pm 4$	11.4
(a) $K^*_0(1950)$	$1942 \pm 22 \pm 5$	$80 \pm 32 \pm 20$	3.3
(b) $K^*_0(1950)$	$1979 \pm 26 \pm 3$	$144 \pm 44 \pm 21$	4.3
$K^*_0(2130)$	$2128 \pm 31 \pm 9$	$95 \pm 42 \pm 76$	2.7
		$\eta_c \rightarrow \eta'\pi^+\pi^-$	
$f_0(500)$	953 ± 90	335 ± 81	
$f_2(1430)$	$1440 \pm 11 \pm 3$	$46 \pm 15 \pm 5$	4.4
$f_0(2100)$	$2116 \pm 27 \pm 17$	$289 \pm 34 \pm 15$	10
		$\eta_c \rightarrow \eta\pi^+\pi^-$	
$a_0(1700)$	$1704 \pm 5 \pm 2$	$110 \pm 15 \pm 11$	8

Table 3.4 Fractions and relative phases from the Dalitz plot analysis of $\eta_c \to \eta' K^+ K^-$

Intermediate state	Fraction (%)	Phase (rad)
$f_0(1710)\eta'$	$29.5 \pm 4.7 \pm 1.6$	0.
$K_0^*(1430)^+ K^-$	$53.9 \pm 7.2 \pm 2.0$	$0.61 \pm 0.13 \pm 0.45$
$K_0^*(1950)^+ K^-$	$2.4 \pm 1.2 \pm 0.4$	$0.46 \pm 0.29 \pm 0.50$
$f_0(1500)\eta'$	$0.8 \pm 1.0 \pm 0.3$	$0.32 \pm 0.54 \pm 0.10$
$f_0(980)\eta'$	$4.7 \pm 2.7 \pm 0.4$	$-0.74 \pm 0.55 \pm 0.05$
$f_2(1270)\eta'$	$2.9 \pm 1.5 \pm 0.11$	$2.9 \pm 0.38 \pm 0.09$
Sum	$94.3 \pm 9.3 \pm 2.6$	
p-value	18%	

An inspection of Fig. 3.7c suggests an additional enhancement in the $m(\eta' K^{\pm})$ around a mass of $\approx$2100 MeV/c^2. We explore this possibility adding, in the Dalitz plot analysis, an additional scalar resonance in this mass region with free parameters. The presence of this additional resonance also affects the parameters of the $K_0^*(1950)$ which are also left free in the fit. The results from this solution are listed in Table 3.3, labelled as solution (b). A comparison between the two fits on the $m(\eta' K^{\pm})$ projection is shown in Fig. 3.7c. However, an application of the Wilks theorem for the individual significances of the K_0^*(1950 and $K_0^*(2130)$ in this new fit, obtain values of 4.3σ and 2.7σ, respectively. Since the local significance of the $K_0^*(2130)$ is less than 3σ, we do not consider the presence of this contribution in the reference fit.

3.3.3.2 Dalitz Plot Analysis of $\eta_c \to \eta' \pi^+ \pi^-$

Figure 3.9a shows the Dalitz plot for the selected $\eta_c \to \eta' \pi^+ \pi^-$ candidates in the data, in the η_c signal region, for the two η' decay modes combined (3122 events), and Fig. 3.9b, c show the two background subtracted projections in linear mass scale. We observe several diagonal bands in the Dalitz plot, in particular at the lower-left edge. There are corresponding structures in the $m(\pi^+\pi^-)$ spectrum, including peaks attributable to the $f_0(980)$ and $f_2(1270)$ resonances, and a large structure at high $\pi^+\pi^-$ mass. A candidate for the large structure in the high $\pi^+\pi^-$ mass region is the $f_0(2100)$ resonance, observed in radiative J/ψ decay to $\gamma\eta\eta$ [7]. We take $f_0(2100)\eta'$ as the reference contribution, and perform a Dalitz plot analysis whose results are given in Table 3.5. We leave free the $f_0(2100)$ resonance parameters and obtain the values reported in Table 3.3 with a significance of 10σ. To describe the small enhancement around 1.43 GeV/c^2, we test both spin-2 and spin-0 hypotheses with free resonance parameters; we obtain $\Delta(-2\log\mathcal{L}) = 2.4$ in favor of the spin-2 hypothesis, so we attribute this signal to the $f_2(1430)$ resonance, and report the fitted parameter values in Table 3.3. We test the significance of this signal by removing it from the list of the resonances, obtaining $\Delta(-2\log\mathcal{L}) = 23.8$ and a significance of 4.4σ.

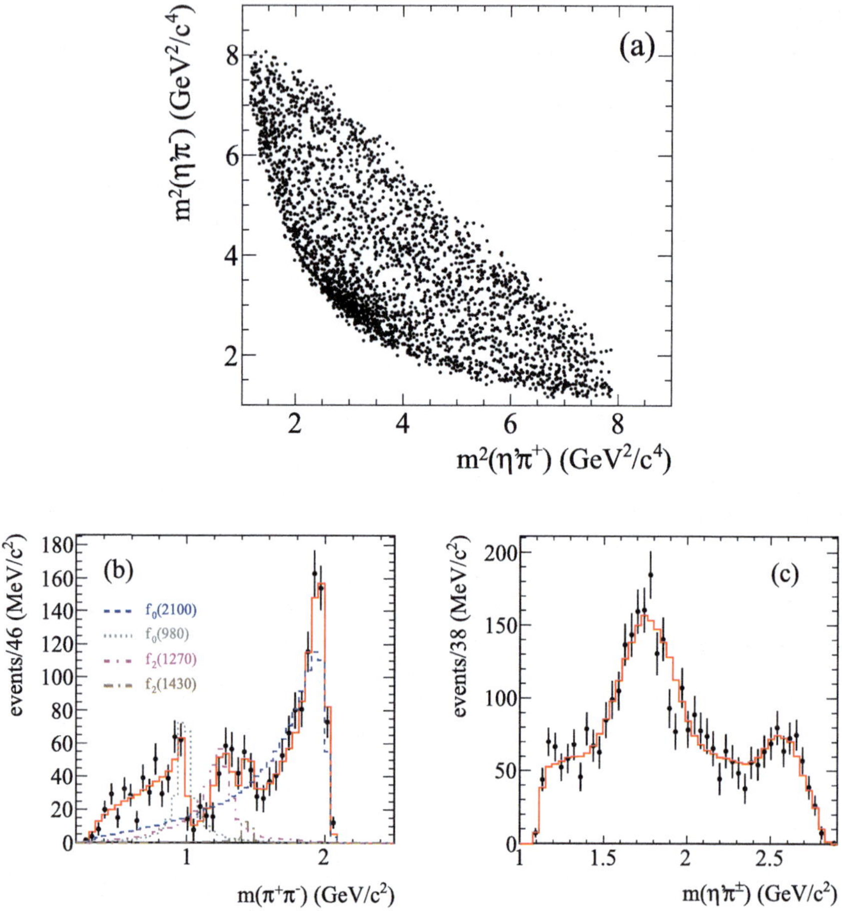

Fig. 3.9 **a** Dalitz plot for selected $\eta_c \to \eta'\pi^+\pi^-$ candidates in the η_c signal region, summed over the two η' decay modes. Linear-scale mass projections **b** $m(\pi^+\pi^-)$ and **c** $m(\eta'\pi^\pm)$, after subtraction of the background. The solid (red) histograms represent the results of the fit described in the text, and the other histograms display the contributions from each of the listed components

3.3.3.3 Dalitz Plot Analysis of $\eta_c \to \eta\pi^+\pi^-$

Figure 3.10a shows the Dalitz plot for the selected $\eta_c \to \eta\pi^+\pi^-$ candidates in the data, in the η_c signal region, for the two η decay modes combined (9303 events), and Fig. 3.10b, c show two background subtracted linear-mass projections. We observe that the Dalitz plot is dominated by horizontal and vertical bands due to the $a_0(980)$ and diagonal bands in the $\pi^+\pi^-$ final state corresponding to $f_0(500)$, $f_0(980)$, and $f_2(1270)$ resonances. We take $a_0(980)^+\pi^-$ as the reference contribution, and perform a Dalitz plot analysis as described above. The resulting list of contributions to this η_c decay mode is given in Table 3.6, together with fitted fractions and relative

Table 3.5 Fractions and relative phases from the Dalitz plot analysis of $\eta_c \to \eta'\pi^+\pi^-$

Intermediate state	Fraction (%)	Phase (rad)
$f_0(2100)\eta'$	$74.9 \pm 7.5 \pm 3.6$	0.
$f_0(500)\eta'$	$4.3 \pm 2.3 \pm 0.7$	$-5.89 \pm 0.24 \pm 0.10$
$f_0(980)\eta'$	$16.1 \pm 2.4 \pm 0.5$	$-5.31 \pm 0.16 \pm 0.04$
$f_2(1270)\eta'$	$22.1 \pm 2.9 \pm 2.4$	$-3.60 \pm 0.16 \pm 0.03$
$f_2(1430)\eta'$	$1.9 \pm 0.7 \pm 0.1$	$-2.45 \pm 0.32 \pm 0.11$
$a_2(1710)\pi$	$3.2 \pm 1.9 \pm 0.5$	$-0.75 \pm 0.27 \pm 0.11$
$a_0(1950)\pi$	$2.5 \pm 1.1 \pm 0.1$	$-0.02 \pm 0.32 \pm 0.06$
$f_2(1800)\eta'$	$5.3 \pm 2.2 \pm 1.4$	$0.67 \pm 0.24 \pm 0.08$
sum	$130.5 \pm 9.5 \pm 4.7$	
p-value	20%	

phases. A new $a_0(1700)$ resonance is observed in the $\eta\pi^\pm$ invariant-mass spectrum, with fitted parameters listed in Table 3.3. The likelihood change obtained when the resonance is excluded from the fit is $\Delta(-2\log\mathcal{L}) = 72.3$, corresponding to a significance greater than 8σ. We note the presence of a very large non-resonant scalar contribution, and in Table 3.6, we list both the sum of resonant contributions and the sum including the non-resonant contribution. This effect could be correlated with the interference of the η_c with the two-photon continuum.

3.3.3.4 Results from the $\eta_c \to \eta K^+K^-$ Analysis

To complete the list of the results summarized in the present review, we also include in Fig. 3.11(Left), the $\eta_c \to \eta K^+K^-$ mass spectrum combined for the $\eta \to \gamma\gamma$ and $\eta \to \pi^+\pi^-\pi^0$ decay modes, first observed by BaBar [26].

Figure 3.11(Right) shows the squared K^+K^- mass projection from the η_c Dalitz plot, where signals of $f_0(1500)$ and $f_0(1700)$ can be seen. The Dalitz plot analysis allow to measure the fractions relative to these resonant contributions which are listed in Table 3.7.

3.4 Conclusions

The study of radiative $\Upsilon(1S)$ decay to $\gamma\pi^+\pi^-$ and γK^+K^- shows the presence of the gluonium candidates $f_0(1500)$ and $f_0(1700)$, in agreement with what observed in J/ψ radiative decays.

In the framework of the identification of scalar gluonium states, it is interesting to compare the rates of η_c decays into a gluonium candidate state and an η or an η' meson.

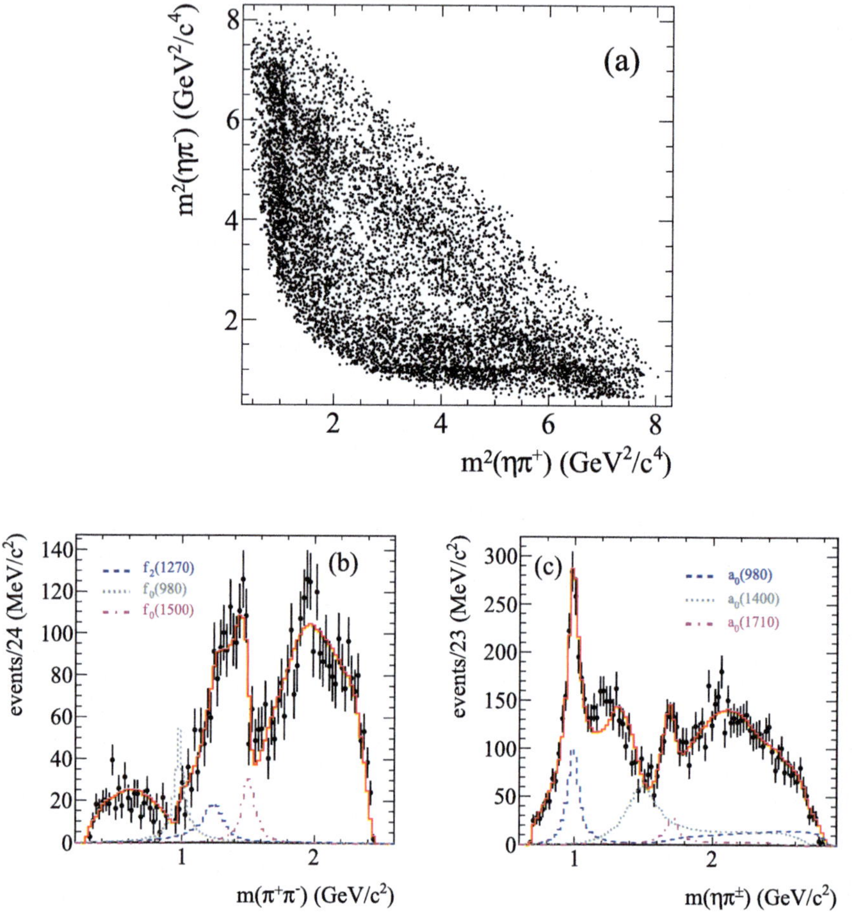

Fig. 3.10 **a** Dalitz plot for selected $\eta_c \to \eta\pi^+\pi^-$ candidates in the η_c signal region, summed over the two η decay modes. Linear-scale mass projections **b** $m(\pi^+\pi^-)$ and **c** $m(\eta\pi^\pm)$, after subtraction of the background. The solid (red) histograms represent the results of the fit described in the text, and the other histograms display the contributions from each of the listed components

Table 3.7 summarizes relevant results from the analyses reported in the present review. We observe an enhanced contribution of $f_0(1710)$ in η_c decays to η' and an enhanced contribution of $f_0(1500)$ in η_c decays to η. This effect may point to an enhanced gluonium content in the $f_0(1710)$ meson. A similar effect is observed for the $f_0(2100)$ resonance. The observation of $f_0(2100)$ in both J/ψ radiative decays and in $\eta_c \to \eta'\pi^+\pi^-$ allows to add this state in the list of the candidates for the scalar glueball.

Table 3.6 Fractions and relative phases from the Dalitz plot analysis of $\eta_c \to \eta\pi^+\pi^-$. The first errors are statistical, the second systematic

Intermediate state	Fraction (%)	Phase (rad)
$a_0(980)^+\pi^-$	$12.3 \pm 1.2 \pm 2.8$	0.
$a_2(1310)^+\pi^-$	$2.5 \pm 0.7 \pm 0.9$	$-1.04 \pm 0.13 \pm 0.20$
$f_0(500)\eta$	$4.3 \pm 1.3 \pm 1.1$	$0.54 \pm 0.14 \pm 0.24$
$f_2(1270)\eta$	$4.6 \pm 0.9 \pm 0.8$	$-1.15 \pm 0.11 \pm 0.05$
$f_0(980)\eta$	$5.7 \pm 1.3 \pm 1.5$	$-2.41 \pm 0.09 \pm 0.07$
$f_0(1500)\eta$	$4.2 \pm 0.7 \pm 0.9$	$2.32 \pm 0.13 \pm 0.17$
$a_0(1450)^+\pi^-$	$15.0 \pm 2.4 \pm 3.2$	$2.60 \pm 0.09 \pm 0.11$
$a_0(1700)^+\pi^-$	$3.5 \pm 0.8 \pm 0.8$	$1.39 \pm 0.15 \pm 0.20$
$f_2(1950)\eta$	$4.2 \pm 1.0 \pm 1.0$	$-1.59 \pm 0.15 \pm 0.21$
Resonant sum	$56.3 \pm 3.7 \pm 10.0$	
NR	$172.7 \pm 8.0 \pm 10.0$	$1.67 \pm 0.07 \pm 0.06$
Sum	$229.0 \pm 8.8 \pm 14.1$	
p-value	9.3%	

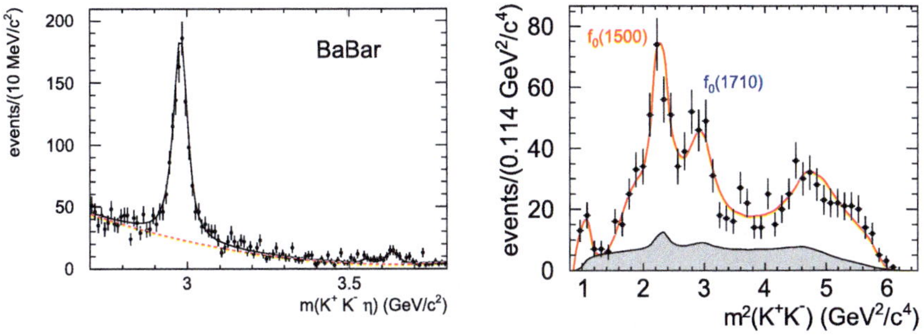

Fig. 3.11 (Left) Invariant ηK^+K^- mass spectrum from $\gamma\gamma \to \eta K^+K^-$. (Right) Squared K^+K^- mass projection from the η_c Dalitz plot. The line is the result from the Dalitz plot analysis

Table 3.7 Fractional contributions to $\eta_c \to \eta h^+h^-$ and $\eta_c \to \eta' h^+h^-$ decays of selected scalar mesons, uncorrected for unseen decay modes

Final state	$f_0(1500)$(%)	$f_0(1710)$(%)	$f_0(2100)$(%)
ηK^+K^-	$23.7 \pm 7.0 \pm 1.8$	$8.9 \pm 0.2 \pm 0.4$	–
$\eta\pi^+\pi^-$	$4.2 \pm 0.7 \pm 0.9$	–	0.
$\eta' K^+K^-$	$0.8 \pm 1.0 \pm 0.3$	$29.5 \pm 4.7 \pm 1.6$	–
$\eta'\pi^+\pi^-$	0.3 ± 0.2	–	$74.9 \pm 7.5 \pm 3.5$

References

1. Y. Chen et al., Phys. Rev. D **73**, 014516 (2006). arXiv:hep-lat/0510074
2. P. Minkowski, W. Ochs, Eur. Phys. J. C **9**, 283–312 (1999). arXiv:hep-ph/9811518
3. C. Amsler, F.E. Close, Phys. Lett. B **353**, 385–390 (1995). arXiv:hep-ph/9505219
4. C. Amsler, F.E. Close, Phys. Rev. D **53**, 295–311 (1996). arXiv:hep-ph/9507326
5. S. Janowski, F. Giacosa, D.H. Rischke, Phys. Rev. D **90**(11), 114005 (2014). arXiv:1408.4921 [hep-ph]
6. L.C. Gui et al., [CLQCD Collaboration], Phys. Rev. Lett. **110**(2), 021601 (2013). arXiv:1206.0125 [hep-lat]
7. M. Ablikim, et al., [BESIII Collaboration], Phys. Rev. D **87**(9), 092009 (2013) [erratum: Phys. Rev. D **87**(11), 119901 (2013)]. arXiv:1301.0053 [hep-ex]
8. L. Kopke, N. Wermes, Phys. Rept. **174**, 67 (1989)
9. S. Dobbs, A. Tomaradze, T. Xiao, K.K. Seth, Phys. Rev. D **91**(5), 052006 (2015). arXiv:1502.01686 [hep-ex]
10. J.P. Lees et al., [BaBar Collaboration], Phys. Rev. D **97**(11), 112006 (2018). arXiv:1804.04044 [hep-ex]
11. S.B. Athar et al., [CLEO Collaboration], Phys. Rev. D **73**, 032001 (2006). arXiv:hep-ex/0510015
12. J.P. Lees et al., [BaBar Collaboration], Phys. Rev. D **104**(7), 072002 (2021). arXiv:2106.05157 [hep-ex]
13. C.N. Yang, Phys. Rev. **77**, 242–245 (1950)
14. L.A. Harland-Lang, V.A. Khoze, M.G. Ryskin, W.J. Stirling, Eur. Phys. J. C **73**, 2429 (2013). arXiv:1302.2004 [hep-ph]
15. S.D. Bass, P. Moskal, Rev. Mod. Phys. **91**(1), 015003 (2019). arXiv:1810.12290 [hep-ph]
16. S.M. Flatte, Phys. Lett. B **63**, 224–227 (1976)
17. T.A. Armstrong et al., [WA76 Collaboration], Z. Phys. C **51**, 351–364 (1991)
18. C. Patrignani et al., [Particle Data Group], Chin. Phys. C **40**, 100001 (2016)
19. J. Becker et al., [Mark-III Collaboration], Phys. Rev. D **35**, 2077 (1987)
20. J.P. Lees et al., [BaBar Collaboration], Phys. Rev. D **85**, 112010 (2012). arXiv:1201.5897 [hep-ex]
21. A.J. Bevan et al., [BaBar and Belle Collaborations], Eur. Phys. J. C **74**, 3026 (2014). arXiv:1406.6311 [hep-ex]
22. P.A. Zyla et al., [Particle Data Group], Prog. Theor. Exp. Phys. **2020**, 083C01 (2020)
23. C.C. Zhang et al., [Belle Collaboration], Phys. Rev. D **86**, 052002 (2012). arXiv:1206.5087 [hep-ex]
24. D. Asner, arXiv:hep-ex/0410014
25. J.M. Blatt, V.F. Weisskopf (Springer, 1952). ISBN 978-0-471-08019-0
26. J.P. Lees et al., [BaBar Collaboration], Phys. Rev. D **89**(11), 112004 (2014). arXiv:1403.7051 [hep-ex]
27. M. Ablikim et al., [BESIII Collaboration], Phys. Rev. D **89**(7), 074030 (2014). arXiv:1402.2023 [hep-ex]
28. J.P. Lees et al., [BaBar Collaboration], Phys. Rev. D **93**, 012005 (2016). arXiv:1511.02310 [hep-ex]
29. D. Aston et al., [LASS Collaboration], Nucl. Phys. B **296**, 493–526 (1988)
30. S.S. Wilks, Ann. Math. Stat. **9**, 60 (1938)

Chapter 4
Global Neutrino Data Analyses

Ivan Martinez-Soler

Abstract The description of neutrino evolution has been a long-standing problem since many decades ago. Considering the three-neutrino mixing scenario, where six parameters describe the neutrino evolution, we study the recent results from a global neutrino analysis that includes the latest results of the most relevant experiments. That global analysis contains data from nuclear reactors, solar neutrinos, accelerators, and atmospheric neutrinos. Combining all those data sets indicates that most of the six parameters are known at the percent level. Still, there are significant uncertainties in the neutrino mass ordering and the leptonic CP-violation phase, which are the primary goal of these analyses.

4.1 Introduction

While neutrinos are massless particles in the Standard Model (SM), the observation of a neutrino flavor oscillation in experiments covering a wide range of energies and baselines proves neutrinos are massive particles. The simplest extension of the SM that accounts for the neutrino masses considers three right-handed neutrinos that are singlets under the SM symmetry group. The mixing between the flavor and the massive states is given by the $3 \times n$ leptonic mixing matrix (U), where n is the number of massive neutrinos. In the case of three massive states, the mixing matrix is parametrized in terms of three mixing angles (θ_{12}, θ_{13}, θ_{23}) and a complex phase that accounts for the violation of the CP-symmetry in the leptonic sector (δ_{CP}). We will use the following parametrization for the lepton mixing matrix

$$U = \begin{pmatrix} 1 & 0 & 0 \\ 0 & c_{23} & s_{23} \\ 0 & -s_{23} & c_{23} \end{pmatrix} \begin{pmatrix} c_{13} & 0 & s_{13}e^{-\delta_{cp}} \\ 0 & 1 & 0 \\ -s_{13}e^{\delta_{cp}} & 0 & c_{13} \end{pmatrix} \begin{pmatrix} c_{12} & s_{12} & 0 \\ -s_{12} & c_{12} & 0 \\ 0 & 0 & 1 \end{pmatrix} \tag{4.1}$$

I. Martinez-Soler (✉)
Department of Physics and Laboratory for Particle Physics and Cosmology, Harvard University, Cambridge, MA 02138, USA
e-mail: imartinezsoler@fas.harvard.edu

G. Ricciardi et al. (eds.), *8th Workshop on Theory, Phenomenology and Experiments in Flavour Physics*, Springer Proceedings in Physics 292,
https://doi.org/10.1007/978-3-031-30459-0_4

Solving the Schrödinger equation, we find that the neutrino flavor oscillates with the ratio between the neutrino energy (E_ν) and the distance traveled (L). In the case of three massive neutrinos, there are two oscillation lengths[1] given by $L_{osc} = 4\pi E_\nu/\Delta m^2_{ij}$, where $\Delta m^2_{ij} = m_i^2 - m_j^2$. The neutrino evolution also depends on the mass hierarchy. Solar experiments have shown that $\Delta m^2_{21} > 0$, but it still is possible to identify two possible orderings depending on whether $\Delta m^2_{31} > 0$ (normal) or $\Delta m^2_{31} < 0$ (inverted).

To determine the oscillation parameters, we need to measure the neutrino evolution at various energies and baselines due to the different energy scales where the oscillations occur. A global fit aims to combine all the experiments where the neutrino flavor oscillations are relevant to obtain a complete description of the neutrino evolution. In the next sections, we will describe the main ingredients, their results, and the tension among them. We will discuss the latest results from NuFit [1], although similar conclusions are also obtained in [2, 3].

4.2 Medium Baseline Reactors

In nuclear reactors, a large flux of $\overline{\nu}_e$ at the MeV is created by the fission of four isotopes, ^{235}U, ^{238}U, ^{239}Pu, and ^{241}Pu. For baselines at the km scale, those neutrinos are sensitive to a combination of the two largest mass splittings, $\Delta m^2_{ee} = \cos^2\theta_{12}|\Delta m^2_{31}| + \sin^2\theta_{12}|\Delta m^2_{32}|$ [4], by looking for the disappearance of $\overline{\nu}_e$. The amplitude of the oscillation is controlled by $\sin^2 2\theta_{13}$.

$$P(\overline{\nu}_e \to \overline{\nu}_e) \approx 1 - \sin^2 2\theta_{13} \sin^2 \Delta_{ee} \tag{4.2}$$

The most recent experimental configurations use a near detector to measure the flux and the cross-section [5–7]. By a rate-only analysis, those experiments have been able to measure $\sin^2 2\theta_{13} = 0.0856 \pm 0.0029$ [6], while a spectral shape analysis allows the constraining of $\Delta m^2_{ee} = (2.522^{+0.068}_{-0.070}) \times 10^{-3}\text{eV}^2$ [6].

Previous theoretical evaluations of the reactor anti-neutrino flux showed several disagreements with the flux measured. In particular, those calculations indicate a 5% deficit in the normalization of the experimental data, the so-called *reactor anti-neutrino anomaly*, and an excess at ~5 MeV over the prediction independent of the reactor power and the distance where the flux is measured. New evaluations of the flux, where an improved calculation of the IBD yield has been used, alleviated those tensions [8]. The near-far configuration used by reactor experiments has allowed a very small impact of the flux anomalies in determining the oscillation parameters.

---

[1] In the case of three massive neutrinos, we have three oscillation lengths, one for each Δm^2_{ij}, but the large hierarchy between the neutrino masses, Δm^2_{31} is almost ~30 bigger than Δm^2_{21}, making that $\Delta m^2_{31} \simeq \Delta m^2_{32}$

4.3 Solar Experiments and Long-Baseline Reactors

The thermonuclear fusion reactions in the Sun are responsible for an electron-neutrino flux produced in the solar core [9]. Different mechanism contributes to this flux; some of them produce a neutrino flux with a very characteristic energy, as is the case of neutrinos from ^{7}Be and *pep*; in other processes, the neutrino flux has a broader energy spectrum that extends to ~10 MeV as in the case of ^{8}B and *hep*.

The evolution of the neutrinos through the Sun changes their flavor composition of the flux. For the energies and baselines of the solar neutrinos $\Delta m^2_{31} >> E_\nu/L$. Therefore the electron survival probability can be written as

$$P^{3\nu}(\nu_e \to \nu_e) \approx \cos^2\theta_{13}\cos^2\theta^m_{13} P^{2\nu}_{eff}(\Delta m^2_{21}, \theta_{12}) + \sin^2\theta^m_{13}\sin^2\theta_{13} \tag{4.3}$$

we have considered that there is no mixing between ν_3 and the other two massive states, which is a good approximation for solar neutrinos due to the small value of θ_{13} and the neutrino energies. The electron survival probability for the 2ν mixing scenario is given by

$$P^{2\nu}_{eff}(\Delta m^2_{21}, \theta_{12}) = \frac{1}{2}(1 + \cos\theta^m_{12}\cos\theta_{12}) \tag{4.4}$$

θ^m_{13} and θ^m_{12} are the effective mixing angles at the neutrino production point. From (4.3) and (4.4), we can see that solar neutrinos are mainly sensitive to θ_{12}, and the dependence on Δm^2_{21} comes from the solar matter effects. Solar also has a very small dependence on θ_{13} and, via the matter effects on Δm^2_{31}. The two experiments that have shown a better sensitivity measuring $P(\nu_e \to \nu_e)$ are Super-Kamiokande [10] and SNO [11].

The most precise determination of Δm^2_{21} was done by KamLAND [12], a liquid-scintillator detector that measured the $\overline{\nu}_e$ flux from nuclear reactors from March 2002 to November 2012. The average baseline for those neutrinos is ~180, which allows us to explore $\Delta m^2 \sim 10^{-5}$ eV ^{2}, two orders of magnitude smaller than in the case of medium-baseline reactors. For those baselines, the survival electron probability is given by

$$P^{3\nu}(\nu_e \to \nu_e) = \cos^4_{13}\left(1 - \frac{1}{2}\sin^2(2\theta_{12})\sin^2\frac{\Delta m^2_{21}L}{2E}\right) + \sin^4_{13} \tag{4.5}$$

The matter effects are negligible for reactor neutrinos due to the low energy of that flux and the lower matter densities crossed.

The comparison between the determination of Δm^2_{21} from solar and reactor experiments shows a 2σ tension between them; KamLAND prefers higher values of the solar mass parameter [1]. The origin of the tension comes from the non-observation of the P_{ee} turn-up expected by the LMA-MSW scenario measured by KamLAND.

The Earth matter effects introduce an asymmetry between measuring the flux during the day and at night. The mass parameter measured by KamLAND predicts a smaller asymmetry than the one measured by Super-K.

4.4 Long-Baseline Accelerators and Atmospheric Experiments

The measurement of the atmospheric parameters (Δm^2_{31}, $\sin\theta_{23}$) and the CP-violation phase (δ_{CP}) requires GeV neutrino beams. Long-baseline experiments (LBL) using a pion source are able to constraint the neutrino oscillation parameters searching for the disappearance of muon neutrinos ($\nu_\mu \rightarrow \nu_\mu$) and the appearance of electron neutrinos ($\nu_\mu \rightarrow \nu_e$). In the case of muon-disappearance, the matter effects are suppressed by $\sin^4\theta_{23}$ [13], and the oscillation probability at leading order [4] can be written as

$$P(\nu_\mu \rightarrow \nu_\mu) \approx 1 - 4|U_{\mu 3}|^2(1 - |U_{\mu 3}|^2)\sin^2 \Delta m^2_{\mu\mu}, \tag{4.6}$$

where

$$\Delta m^2_{\mu\mu} = \sin^2\theta_{12}\Delta m^2_{31} + \cos^2\theta_{12}\Delta m^2_{32} + \cos\delta_{CP}\sin\theta_{13}\sin 2\theta_{12}\tan\theta_{23}\Delta m^2_{21} \tag{4.7}$$

The muon disappearance channel is mainly sensitive to $|\Delta m^2_{31}|$ and $\sin^2 2\theta_{23}$ measuring whether θ_{23} is maximal mixing or not. Notice that $P_{\mu\mu}$ and P_{ee} are sensitive to a different combination of Δm^2_{31} and Δm^2_{32}. Therefore combining muon and electron disappearance measurements, we can get sensitivity over the neutrino mass ordering [4].

In the appearance channel, the matter effects are more important, so we can use an expression valid for the neutrino evolution in constant matter [14]

$$P_{\nu_\mu\rightarrow\nu_e} \approx 4\sin^2\theta_{13}\sin^2\theta_{23}(1 + 2oA) - C\sin\delta_{cp}(1 + oA) \tag{4.8}$$

where o is the sign of Δm^2_{31}, and $C = \Delta m^2_{21}L/4E\sin 2\theta_{12}\sin 2\theta_{13}\sin 2\theta_{23}$. The matter effects are introduced by $A = 2EV/\Delta m^2_{31}$, where V is the matter potential. The sensitivity to the neutrino mass ordering in the electron appearance channel comes from the matter effects. The dependence on $\sin\theta_{23}$ brings the possibility of separating between the lower and the higher octant. This channel is also sensitive to the CP-phase. The possibility of those experiments running in the neutrino and the anti-neutrino mode increase the sensitivity over δ_{CP}.

The most relevant LBL experiments are T2K and NOvA, which show tension in the determination of δ_{CP} [15] and the mass ordering (NO). For normal ordering, NOvA has a preference for $\delta_{CP} \sim \pi/2$, whereas T2K prefers values around $\delta_{CP} \sim 3\pi/2$. For invert ordering (IO), the region allowed by both experiments coincide with $\delta_{CP} \sim 3\pi/2$. The combined analysis of both experiments shows a preference for IO due to the disagreement in δ_{CP} and NO. Regarding Δm^2_{31}, the measurement is dominated by T2K although both experiments show a good agreement. For $\sin^2\theta_{23}$, there is a preference from both experiments for higher octant, and maximal mixing is excluded to less than 2σ.

In the atmosphere, a flux of ν_e and ν_μ is created by the collision of cosmic rays with the atmospheric nuclei and the subsequent decay of secondary meson flux. The atmospheric neutrino flux extends from ~10 MeV to ~100 TeV, and its measurement below the GeV scale provides sensitivity over the CP-violation phase due to the development of the Δm^2_{21} oscillation and the CPT conservation [16–18]. Around ~6 GeV and for the trajectories crossing the mantle, neutrinos undergo a flavor resonance that happens for neutrinos if $\Delta m^2_{31} > 0$ or for anti-neutrino if $\Delta m^2_{31} < 0$. The differences in flux and cross section between neutrinos and anti-neutrinos allow resolving the ordering in the atmospheric neutrino flux. Neutrino have the first oscillation minimum for baselines crossing Earth at energies ~20 GeV. The location of the minimum is determined by Δm^2_{31}, and the amplitude of the oscillation by $\sin^2 2\theta_{23}$.

The most relevant atmospheric neutrino experiments are IceCube/DeepCore [19], which measures the multi-GeV part of the flux, and Super-Kamiokande [20], which is sensitive to neutrinos with energies above ~500 MeV. By the moment, atmospheric neutrino results are consistent with LBL, although they show a lower precision on Δm^2_{31} and $\sin^2\theta_{23}$. Regarding the neutrino mass ordering, Super-Kamiokande shows a $\sim 2\sigma$ preference for NO, and for the CP-phase, just the region around $\delta_{CP} \sim \pi/4$ is excluded at $\sim 2\sigma$ for both mass orderings.

4.5 Conclusions

The 3-neutrino mixing scenario explains with good accuracy most of the data measured in reactor, accelerator, solar and atmospheric data. In Fig. 4.1, we show the sensitivity to each oscillation parameter from the latest NuFit results [1]. Among the less constrained parameters, we found $\sin^2_{23}$, the neutrino mass ordering, and δ_{CP}.

The latest results from LBL indicate a preference for IO due to the tension in determining δ_{CP}. If LBL are combined with reactor experiments, NO is preferred with a significance lower than 2σ. If those results are combined with the latest results of Super-kamiokande, the preference for NO rises to $\sim 2.7\sigma$. Regarding $\sin^2\theta_{23}$, both octants are allowed, although there is a preference for higher octant from all the LBL experiments, and the maximal mixing is excluded with less than 2σ. For the CP-violation phase, the results are dominated by T2K, and the prefer region is around $\delta_{CP} \sim \pi$ for normal ordering. For invert ordering all the experiments coincide with the preference of $\delta_{CP} \sim 3\pi/2$.

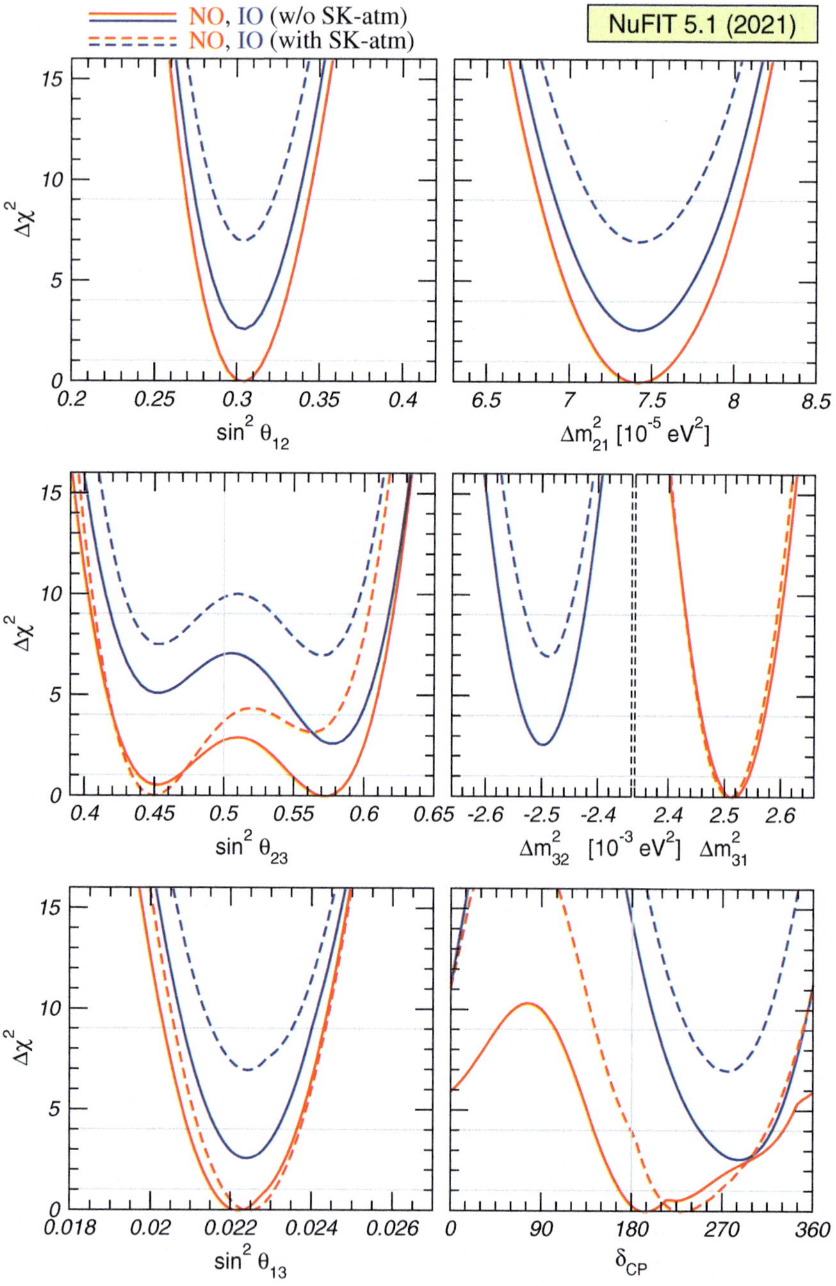

Fig. 4.1 Global neutrino analysis including the most updated data until October 2021

References

1. I. Esteban, M.C. Gonzalez-Garcia, M. Maltoni, T. Schwetz, A. Zhou, JHEP **09**, 178 (2020). https://doi.org/10.1007/JHEP09(2020)178. arXiv:2007.14792 [hep-ph]
2. F. Capozzi, E. Di Valentino, E. Lisi, A. Marrone, A. Melchiorri, A. Palazzo, Phys. Rev. D **104**(8), 083031 (2021). https://doi.org/10.1103/PhysRevD.104.083031. arXiv:2107.00532 [hep-ph]
3. P.F. de Salas, D.V. Forero, S. Gariazzo, P. Martínez-Miravé, O. Mena, C.A. Ternes, M. Tórtola, J.W.F. Valle, JHEP **02**, 071 (2021). https://doi.org/10.1007/JHEP02(2021)071. arXiv:2006.11237 [hep-ph]
4. H. Nunokawa, S.J. Parke, R. Zukanovich Funchal, Phys. Rev. D **72**, 013009 (2005). https://doi.org/10.1103/PhysRevD.72.013009. arXiv:hep-ph/0503283 [hep-ph]
5. G. Bak et al., [RENO], Phys. Rev. Lett. **121**(20), 201801 (2018). https://doi.org/10.1103/PhysRevLett.121.201801. arXiv:1806.00248 [hep-ex]
6. D. Adey et al., [Daya Bay], Phys. Rev. Lett. **121**(24), 241805 (2018). https://doi.org/10.1103/PhysRevLett.121.241805. arXiv:1809.02261 [hep-ex]
7. H. de Kerret et al., [Double Chooz], Nature Phys. **16**(5), 558–564 (2020). https://doi.org/10.1038/s41567-020-0831-y. arXiv:1901.09445 [hep-ex]
8. C. Giunti, Y.F. Li, C.A. Ternes, Z. Xin, Phys. Lett. B **829**, 137054 (2022). https://doi.org/10.1016/j.physletb.2022.137054. arXiv:2110.06820 [hep-ph]
9. R.L. Workman et al., [Particle Data Group], PTEP **2022**, 083C01 (2022). https://doi.org/10.1093/ptep/ptac097
10. K. Abe et al., [Super-Kamiokande], Phys. Rev. D **94**(5), 052010 (2016). https://doi.org/10.1103/PhysRevD.94.052010. arXiv:1606.07538 [hep-ex]
11. B. Aharmim et al., SNO. Phys. Rev. C **88**, 025501 (2013). https://doi.org/10.1103/PhysRevC.88.025501. arXiv:1109.0763 [nucl-ex]
12. A. Gando et al., [KamLAND], Phys. Rev. D **88**(3), 033001 (2013). https://doi.org/10.1103/PhysRevD.88.033001. arXiv:1303.4667 [hep-ex]
13. E.K. Akhmedov, M. Maltoni, A.Y. Smirnov, JHEP **05**, 077 (2007). https://doi.org/10.1088/1126-6708/2007/05/077. arXiv:hep-ph/0612285 [hep-ph]
14. J. Elevant, T. Schwetz, JHEP **09**, 016 (2015). https://doi.org/10.1007/JHEP09(2015)016. arXiv:1506.07685 [hep-ph]
15. K.J. Kelly, P.A.N. Machado, S.J. Parke, Y.F. Perez-Gonzalez, R.Z. Funchal, Phys. Rev. D **103**(1), 013004 (2021). https://doi.org/10.1103/PhysRevD.103.013004. arXiv:2007.08526 [hep-ph]
16. M. Jiang et al., [Super-Kamiokande], PTEP **2019**(5), 053F01 (2019). https://doi.org/10.1093/ptep/ptz015. arXiv:1901.03230 [hep-ex]
17. K.J. Kelly, P.A. Machado, I. Martinez Soler, S.J. Parke, Y.F. Perez Gonzalez, Phys. Rev. Lett. **123**(8), 081801 (2019). https://doi.org/10.1103/PhysRevLett.123.081801. arXiv:1904.02751 [hep-ph]
18. I. Martinez-Soler, H. Minakata, PTEP **2019**(7), 073B07 (2019). https://doi.org/10.1093/ptep/ptz067. arXiv:1904.07853 [hep-ph]
19. T. Stuttard, Particle physics with atmospheric neutrinos at IceCube/DeepCore. Neutrino (2022). https://doi.org/10.5281/ZENODO.6694972. https://zenodo.org/record/6694972
20. L. Wan, New results with atmospheric neutrinos at Super-Kamiokande. Neutrino (2022). https://doi.org/10.5281/zenodo.6694761

Chapter 5
The SND@LHC Experiment

Giovanni De Lellis

Abstract SND@LHC, Scattering and Neutrino Detector at the LHC, is a compact experiment designed to perform measurements with neutrinos produced at the LHC in the unexplored pseudo-rapidity region of $7.2 < \eta < 8.4$, complementary to all the other experiments at the LHC. The experiment was approved in March 2021. It was constructed in about one year and it is now taking data during the Run 3 of the LHC. In this paper we review the detector concept, the physics case and the status of the data taking.

5.1 Introduction

As the accelerator with the highest beam energy, the LHC is also the source of the most energetic human-made neutrinos. Indeed, the LHC produces an intense and strongly collimated beam of TeV-energy neutrinos along the direction of the proton beams. Notably, this neutrino beam includes a sizable fraction of tau neutrinos, mainly produced via the $D_s \to \tau\nu_\tau$ decay and subsequent τ decays, and hence provides a novel opportunity to study their properties.

Already in 1984, De Rujula and Rückl proposed to use the LHC neutrino beam by placing a neutrino experiment in the far forward direction [1]. This idea of detecting LHC neutrinos was revisited several times in the following decades [2, 3]. More recently, a feasibility study was carried out, resulting in the estimate of the physics potential and in the identification of a proper location underground in the LHC tunnel for such an experiment to operate during the Run3 of the LHC [4, 5]. In 2018, the FASER collaboration installed a suitcase size pilot detector employing emulsion films and recently reported the first neutrino interaction candidates at the LHC [6].

G. De Lellis (✉)
Dipartimento di Fisica E. Pancini, Università di Napoli Federico II,
via Cintia 19, 80126 Napoli, Italy
e-mail: giovanni.de.lellis@cern.ch

Istituto Nazionale di Fisica Nucleare, via Cintia 19, 80126 Napoli, Italy

G. Ricciardi et al. (eds.), *8th Workshop on Theory, Phenomenology and Experiments in Flavour Physics*, Springer Proceedings in Physics 292,
https://doi.org/10.1007/978-3-031-30459-0_5

The SND@LHC Collaboration submitted a Letter of Intent in August 2020 [7] and a Technical Proposal in January 2021 [8].

5.2 Experiment Concept

The experiment is located 480 m downstream of IP1 in the TI18 tunnel, an injection tunnel during LEP operation. The detector consists of a hybrid system based on an 830 kg target mass of tungsten plates, interleaved with emulsion and electronic trackers, followed downstream by an hadronic calorimeter and a muon identification system, as shown in Fig. 5.1. The emulsion films with their micrometric accuracy [9] constitute the vertex detector while trackers in the target region, based on the Scintillating Fibre technology [10], provide the time stamp to the events and complement emulsion for the electromagnetic energy reconstruction. Nuclear emulsion films are readout by state-of-the-art, fully automated, optical scanning systems [11–14]. The hadronic calorimeter and muon system comprises eight layers of scintillating bar planes interleaved with 20 cm-thick iron slabs. The three most downstream stations are made of fine grained bars with both horizontal and vertical orientation, to trace the penetrating muons. Every scintillating bar as well as every fibre module is viewed by SiPMs.

The detector configuration allows efficiently distinguishing between all three neutrino flavours, as well as searching for Feebly Interacting Particles via signatures of scattering in the detector target [15]. The first phase aims at operating the detector throughout Run 3 to collect about 290 fb^{-1}.

The detector takes full advantage of the space available in the TI18 tunnel to cover the desired range in pseudo-rapidity. Figure 5.2 shows the top and side views of the detector positioned inside the tunnel. It is worth noting that the tunnel floor is sloped, as can be seen from the side view, with the floor sloping down along the longitudinal axis of the detector. As shown in the top view, the nominal collision axis from IP1 comes out of the floor very close to the wall of the tunnel. Since no civil engineering work could have been done in time for the operation in Run 3, the tunnel

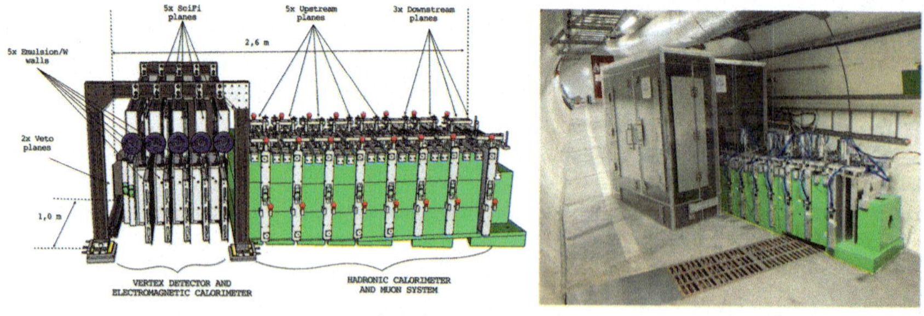

Fig. 5.1 Detector layout (left) and picture of the detector installed in TI18 (right)

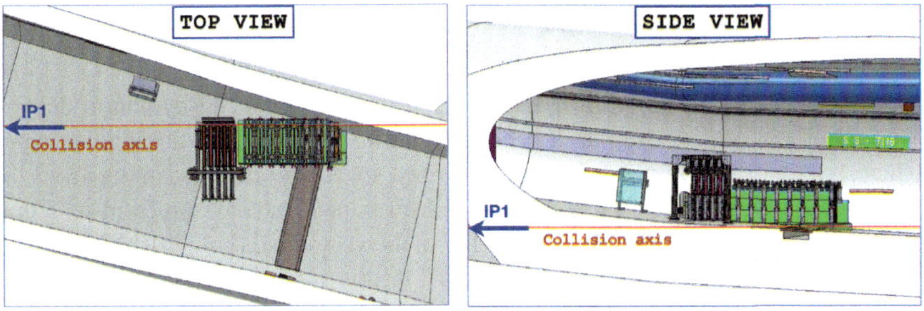

Fig. 5.2 Side and top views of the SND@LHC detector in the TI18 tunnel [8]

geometry imposed several constraints. The following guidelines were adopted for the optimisation of the detector design: a good calorimetric measurement of the energy requires about 10 λ_{int}; a good muon identification efficiency requires enough material to absorb hadrons; for a given transverse size of the target region, the azimuthal angular acceptance decreases with the distance from the beam axis. The energy measurement and the muon identification set a constraint on the minimum length of the detector. With the constraints from the tunnel, this requirement competes with the azimuthal angular acceptance that determines the overall flux intercepted and therefore the total number of observed interactions. The combination of position and size of the proposed detector is an optimal compromise between these competing requirements. The geometrical constraints also restrict the detector to the first quadrant only around the nominal collision axis, as shown in the top view of the detector in Fig. 5.2.

The result is a compact detector, 2.6 m in length. The energy measurement and the muon identification limit the target region to a length of about 80 cm. The transverse size downstream of about 80(H) $\times$ 60(V) cm^2 is limited by the constraint of the tunnel side wall. The transverse size of the target region is proportionally smaller in order to match the acceptance of the energy measurement and the muon identification for the vertices identified in the target volume. In order to maximise the number of neutrino interactions, tungsten has been selected as the passive material. The emulsion target will be replaced a few times per year.

With data from Run 3, SND@LHC will be able to study more than two thousand high-energy neutrino interactions.

All the detector systems were constructed in the labs by Summer 2021 and were assembled and tested at CERN. On November 1st, the installation underground started. A borated polyethylene shielding box was added to surround the target and absorb low-energy neutrons originated from beam-gas interactions as shown in the right picture of Fig. 5.1. The detector installation was completed on April 7th 2022 by adding the target walls with emulsion films, and it is now taking data with the Run 3 of the LHC.

We review in the following the physics case of the experiment.

5.2.1 QCD Measurements

Electron neutrinos in $7.2 < \eta < 8.4$ range are mostly produced by charm decays. Therefore, ν_es can be used as a probe of charm production in an angular range where the charm yield has a large uncertainty, to a large extent coming from the gluon parton distribution function (PDF). Electron neutrino measurements can thus constrain the uncertainty on the gluon PDF in the very small (below 10^{-5}) x region. The interest therein is two-fold: gluon PDF in this x domain will be relevant for Future Circular Collider (FCC) detectors; secondly, the measurement will reduce the uncertainty on the flux of very-high-energy atmospheric neutrinos produced in charm decays, essential for the evidence of neutrinos from astrophysical sources [16, 17]. The charm measurement in Run3 will be affected by a systematic uncertainty at the level of 30% and by a statistical uncertainty of 5%.

The left plot of Fig. 5.3 shows the ratio between charm measurements in different η regions normalised to the LHCb measurement [18]: gluon PDF uncertainty provides the largest contribution. SND@LHC will measure charm in the $7.2 < \eta < 8.4$ region where the PDF uncertainty is dominant.

5.2.2 Lepton Flavour Universality with Neutrino Interactions

In the pseudo-rapidity range of interest, tau neutrinos are essentially only produced in $D_s \to \tau\nu_\tau$ and the subsequent τ decays. One can thus assume that the source of both ν_e and ν_τ is essentially provided by semi-leptonic and fully leptonic decays of charmed hadrons. Unlike ν_τs produced only in D_s decays, ν_es are produced in the decay of all charmed hadrons, essentially D^0, D, D_s and Λ_c. Therefore, the ν_e/ν_τ ratio depends only on the charm hadronisation fractions and decay branching ratios.

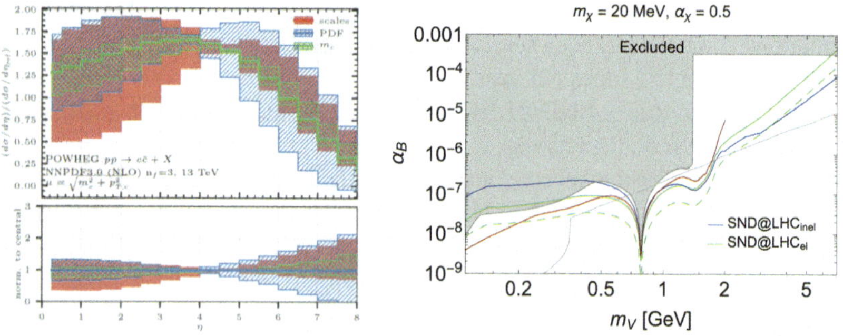

Fig. 5.3 Left: Ratio between the differential cross-section at 13 TeV and the differential cross-section at 7 TeV, with the latter evaluated in the pseudo-rapidity range $4 < \eta < 4.5$ [8]. Right: Sensitivity of the SND@LHC experiment to the leptophobic portal [15]

The systematic uncertainties due to the charm-quark production mechanism cancel out, and the ratio becomes sensitive to the ν-nucleon interaction cross-section ratio of the two neutrino species. The measurement of this ratio can thus be considered a lepton flavour universality test in neutrino interactions.

Charmed hadron fractions and ν branching ratios in the experiment acceptance produce a systematic uncertainty on this ratio of about 22% while the statistical uncertainty is dominated by the low statistics of the ν_τ sample, which corresponds to a 30% accuracy [8]. The systematic uncertainty was evaluated by studying the fluctuations of the ratio using different event generators.

Lepton flavour universality can also be tested with the electron to muon neutrino ratio. The ν_μs are much more abundant but heavily contaminated by π and K decays, and therefore the production mechanism cannot be considered the same as in the case of ν_e. However, this contamination is mostly concentrating at low energies. Above 600 GeV, the contamination is predicted to be reduced to about 35%, and stable with the energy. Moreover, charmed hadron decays have practically equal branching ratios into electron and muon neutrinos. As a result, the ν_e/ν_μ ratio provides a test of the lepton flavour universality with an uncertainty of 15%, with an equal 10% statistical and systematic contribution [8].

5.2.3 *Feebly Interacting Particles*

The experiment is also capable of performing model-independent direct searches for FIPs. They may be produced in the pp scattering at the LHC interaction point, propagate to the detector and decay or scatter inside it. The background from neutrino interactions can be rejected by a time-of-flight measurement.

A recent work [15] summarises the experiment sensitivity to physics beyond the Standard Model, by considering the scatterings of light dark matter particles χ via leptophobic $U(1)_B$ mediator, as well as decays of Heavy Neutral Leptons, dark scalars and dark photons. The excellent spatial resolution of nuclear emulsions makes SND@LHC suited to search for neutral mediators decaying into two charged particles.

SND@LHC is unique in its capability to perform a direct dark matter search at accelerators. The right plot of Fig. 5.3 shows the sensitivity of the experiment to the leptophobic portal under the assumption that $m_\chi = 20$ MeV and the coupling of the mediator to χ particles is $\alpha_\chi = 0.5$. The considered signatures are the elastic scattering off protons (green line, 10 signal events) and the deep-inelastic scattering (blue line, 100 signal events). The dashed line corresponds to the upgraded setup that may operate during Run 4. The red line shows the 100 event contour for the DUNE experiment [19].

5.3 Outlook

The last sixteen months have been a full-fledged race against the clock for the Collaboration. The experiment is now taking data during the Run 3 of the LHC and the full apparatus is being commissioned with collision data. While a few small independent emulsion bricks exposed during the LHC commissioning have already been developed, indicating a negligible background level, the collaboration is starting the analysis of the first set of emulsion from the target region. The extraction of the emulsion and the installation of a fully instrumented detector were successfully done at the end of July, halfway through the LHC intensity ramp-up. In the meantime, data from the electronics detector are continuously analysed. A new era of collider neutrinos has just started.

References

1. A. De Rujula, R. Ruckl, Neutrino and muon physics in the collider mode of future accelerators. CERN-TH-3892/84. https://doi.org/10.5170/CERN-1984-010-V-2.571
2. A. De Rujula et al., Neutrino fluxes at future hadron colliders. Nucl. Phys. B **405**, 80–108 (1993). CERN-TH-6452-92. https://doi.org/10.1016/0550-3213(93)90427-Q
3. H. Park, The estimation of neutrino fluxes produced by proton-proton collisions at $\sqrt{s} = 14$ TeV of the LHC. JHEP **10**, 092 (2011). https://doi.org/10.1007/JHEP10(2011)092
4. S. Buontempo et al., CMS-XSEN: LHC neutrinos at CMS. Experiment feasibility study. arXiv:1804.04413
5. N. Beni et al., Physics potential of an experiment using LHC neutrinos. J. Phys. G **46**, 115008 (2019). https://doi.org/10.1088/1361-6471/ab3f7c
6. H. Abreu et al., [FASER Collaboration], First neutrino interaction candidates at the LHC. Phys. Rev. D **104**, L091101 (2021)
7. SND@LHC Collaboration, Scattering and Neutrino Detector at the LHC, LoI, CERN-LHCC-2020-013, LHCC-I-037. http://cds.cern.ch/record/2729015
8. C. Ahdida et al., SND@LHC - scattering and neutrino detector at the LHC. Technical Proposal, CERN-LHCC-2021-003, LHCC-P-016 (2021). https://cds.cern.ch/record/2750060
9. C. Fabjan, H. Schopper, *Particle Physics Reference Library*. Volume 2: Detectors for Particles and Radiation (Springer Nature). https://doi.org/10.1007/978-3-030-35318-6
10. LHCb Collaboration, LHCb tracker upgrade technical design report, CERN-LHCC-2014-001, LHCB-TDR-015 (2014). http://cds.cern.ch/record/1647400
11. A. Alexandrov et al., A new fast scanning system for the measurement of large angle tracks in nuclear emulsions. JINST **10**, P11006 (2015)
12. A. Alexandrov et al., A new generation scanning system for the high-speed analysis of nuclear emulsions. JINST **11**, P06002 (2016)
13. A. Alexandrov et al., The continuous motion technique for a new generation of scanning systems. Sci. Rep. **7**, 7310 (2017)
14. A. Alexandrov, G. De Lellis, V. Tioukov, A novel optical scanning technique with an inclined focusing plane. Sci. Rep. **9**, 2870 (2019)
15. A. Boyarsky et al., Searches for new physics at SND@LHC. JHEP **03**, 006 (2022). https://doi.org/10.1007/JHEP03(2022)006

16. A. Bhattacharya et al., Prompt atmospheric neutrino fluxes: perturbative QCD models and nuclear effects. JHEP **11**, 167 (2016)
17. Y.S. Jeong, Neutrinos from charm: forward production at the LHC and in the atmosphere. PoS 1218 (2021). https://doi.org/10.22323/1.395.1218
18. R. Aaij et al., [LHCb Collaboration], Measurements of prompt charm production cross-sections in *pp* collisions at $\sqrt{s} = 13$ TeV. JHEP **03**, 159 (2016). https://doi.org/10.1007/JHEP03(2016)159
19. S. Naaz et al., Adv. High Ener. Phys. 1–9 (2020). https://doi.org/10.1155/2020/9047818

Chapter 6
Seeking New Physics at Neutrino Oscillation Experiments

Vedran Brdar

Abstract One of the most important achievements in the field of particle physics was the discovery of neutrino oscillations. Neutrino oscillation experiments still have a lot to offer, primarily the discovery of CP violation in the lepton sector is anticipated. In addition to solving the remaining puzzles in the standard three-neutrino framework, neutrino experiments are also sensitive to new physics effects that could appear in the process of neutrino production, propagation and/or detection. We discuss a novel class of new physics realizations, testable already at present-day acceleration based experiments such as NOvA and T2K. The idea is based on the fact that neutrino mixing parameters at the scale of neutrino production and neutrino detection do not necessarily need to coincide given that they are subject to renormalization group effects. This mismatch between the leptonic mixing matrix at different scales changes the neutrino oscillation phenomenology by inducing a number of new signatures, some of which are discussed here.

6.1 Introduction

Despite the already awarded Nobel Prize, neutrino oscillations still have a lot to offer. In particular, the discovery of CP violation in the lepton sector, mass ordering of neutrinos and the octant of the atmospheric mixing angle is anticipated. In addition to unraveling these remaining unknowns in the standard three-flavor neutrino oscillation paradigm, such experiments also have a great potential for testing new physics. For instance, light sterile neutrinos [1] and non-standard interactions [2] can be probed. In what follows we discuss a novel class of beyond the Standard Model physics that can lead to complementary signatures at neutrino oscillation experiments.

The probability that a neutrino produced as ν_α, after propagating a distance L, gets detected as ν_β, $(\alpha, \beta = e, \mu, \tau)$ reads

V. Brdar (✉)
Theoretical Physics Department, CERN, Esplande des Particules, 1211 Geneva 23, Switzerland
e-mail: vedran.brdar@cern.ch

G. Ricciardi et al. (eds.), *8th Workshop on Theory, Phenomenology and Experiments in Flavour Physics*, Springer Proceedings in Physics 292,
https://doi.org/10.1007/978-3-031-30459-0_6

$$P_{\alpha\beta} = \sum_{j,k} U^*_{\alpha j} U_{\beta j} U_{\alpha k} U^*_{\beta k} \, e^{-i \frac{\Delta m^2_{jk}}{2E_\nu} L} \,, \tag{6.1}$$

where $j, k = 1, 2, 3$, $\Delta m^2_{jk} \equiv m^2_j - m^2_k$ where m_i is the neutrino mass associated to the eigenstate ν_i, and E_ν is the neutrino energy. The leptonic mixing matrix (PMNS) elements with index α (β) are associated to the neutrino production (detection). These processes occur at different energies; e.g. in the acceleration-based experiments, the production scale corresponds to the pion mass while the detection scale, being a function of E_ν and nucleon mass, is about an order of magnitude larger. If significant renormalization group effects can induce the change of PMNS matrix across these scales, Eq. (6.1) would change because there would be essentially two related but different PMNS matrices entering the formula for neutrino oscillations. Let us illustrate that by looking at the two-flavor system in absence of any CP violation. The usual formula for neutrino oscillations reads

$$P_{e\mu} = \sin^2 2\theta \sin^2 \left(\frac{\Delta m^2 L}{4E_\nu} \right) , \tag{6.2}$$

while in the presence of running effects the formula changes to [3]

$$P_{e\mu} = \sin^2(\theta_p - \theta_d) + \sin 2\theta_p \sin 2\theta_d \sin^2 \left(\frac{\Delta m^2 L}{4E_\nu} \right) . \tag{6.3}$$

Here, θ_p and θ_d are mixing angles corresponding to production and detection scale, respectively and, as expected, it is clear that if $\theta_p = \theta_d$ Eq. (6.2) is reproduced. The first term in Eq. (6.3) indicates that zero-baseline flavor transitions occur in the setup with running. Another interesting signature, given that the mixing angle θ changes with the energy scales, would be measuring different values of θ at experiments which operate at different energy scales.

6.2 Phenomenology

In this section we will discuss three complementary phenomenological features that can be induced at neutrino experiments by renormalization group effects.

6.2.1 Neutrino Oscillations at T2K and NOνA

T2K and NOνA can measure electron neutrino appearance from the beam that mostly consists of muon neutrinos. In [3] we have introduced a model, within which neutrino masses are generated at 1-loop level, that features large renormalization group

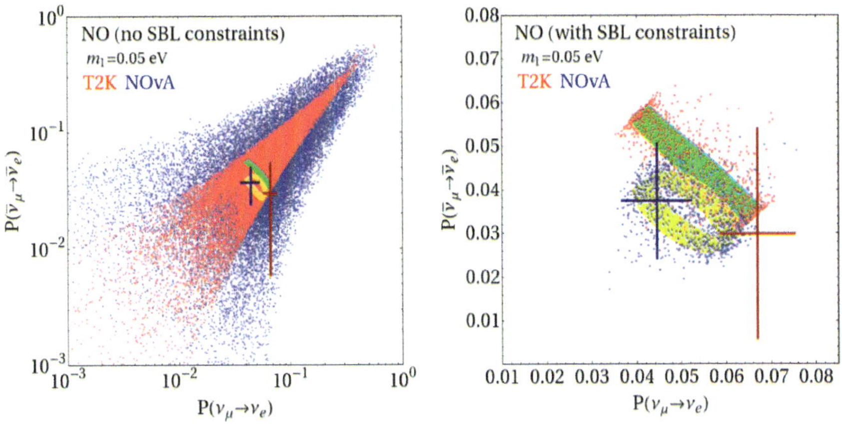

Fig. 6.1 Standard regions (green and yellow), obtained by varying oscillation parameters within 1σ from their best fit points [4], and data from T2K and NOvA are compared to regions that are obtained by evaluating modified three-flavor oscillation probabilities within the model [3] that features large renormalization group effects (red and blue scatter points). In the left panel we illustrate how large the effect can in principle be, while in the right panel we employ constraints from short-baseline experiments and indicate the magnitude that such new physics effects can yield at present experiments

effects across the scales relevant for neutrino production and detection. This leads to the modification of oscillation formulae as we already demonstrated in Eq. (6.3). Numerically, we show this in Fig. 6.1 via red and blue scatter points which represent different choices in the parameter space of the model. In the left panel we illustrate how large the effect can in principle be, while in the right panel we show that such region drastically reduces after applying constraints from short-baseline experiments. Nevertheless, even after applying such constraints, red and blue points exceed the standard regions for NOvA and T2K, shown in yellow and green, respectively.

6.2.2 *IceCube High-Energy Neutrinos*

The renormalization group effects are stronger for larger separation between production and detection scale. The detection scale grows with E_ν and therefore it appears worthwhile to inspect neutrinos of highest energy detected to date by the IceCube collaboration [5]. In doing so, we focus on the flavor composition of high-energy neutrinos, which is a powerful observable for testing new physics. We do not know which mechanism is responsible for the production of neutrinos from extragalactic sources and to this end we explored several different scenarios in [3]. Here, for brevity, we will only focus on the neutron decay mechanism where it is only electron (anti)neutrinos that get produced at the source. Such production mechanism is already disfavored by the data; in particular, the green region in Fig. 6.2 lies out-

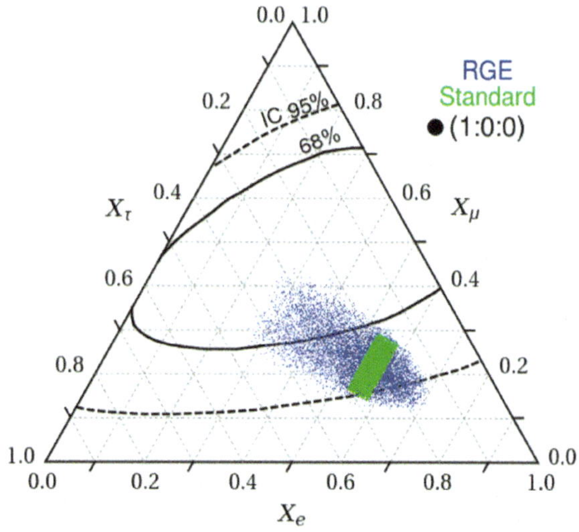

Fig. 6.2 Flavor composition of astrophysical neutrinos at Earth. In green (blue) we show the accessible regions without and with new physics, respectively. One can observe that the fraction of the scatter points corresponding to the running effect scenario (blue) lies within 68% CL region obtained by IceCube. This makes the running effect relevant already in light of present data since such new physics strengthens the case for the neutron decay as the production mechanism of high-energy neutrinos

side of 68% CL region constrained by IceCube. However, the situation changes in the framework with energy-dependent mixing angles. After performing a parameter scan in the model, we obtained blue scatter points, some of which clearly lie within 68% CL region set by IceCube. We conclude that in the presence of this type of new physics, the consistency of neutron decay production mechanism in light of the present data increases.

6.2.3 Sterile Neutrinos

The MiniBooNE collaboration has announced the 4.8σ excess of electron-like events [6] which hints neutrino oscillations at very short baseline and this is not consistent with the standard three-flavor neutrino oscillation paradigm. While the MiniBooNE anomaly got alleviated in recent studies [7, 8], it still stands at $\sim 3\sigma$. This calls for the introduction of new physics and the simplest scenario is consideration of eV-scale sterile neutrino that induces oscillations across short baselines and hence explains the anomaly. However, the preferred parameter space for such a model is in strong tension with the muon neutrino disappearance searches at IceCube and MINOS [9].

Let us demonstrate that by taking eV-scale sterile neutrino and assuming that its mixing angles grow with energy via renormalization group effects (see [10] for

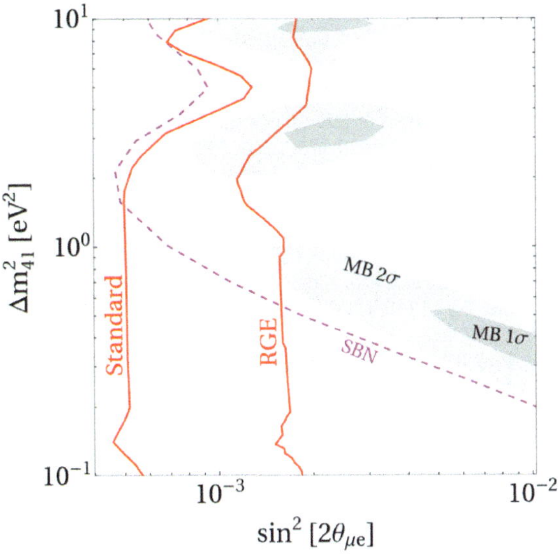

Fig. 6.3 'Standard:' Combined constraints on $\sin^2 2\theta_{\mu e}$ as a function of Δm^2_{41}, assuming the standard scenario with eV-scale sterile neutrino. 'RGE:' Combined constraints on the effective $\sin^2 2\theta_{\mu e}$ at MiniBooNE as a function of Δm^2_{41} in the model with running effects. The region of parameter space preferred by MiniBooNE is depicted in grey. The dashed, purple line is the 3σ CL sensitivity of Fermilab's Short Baseline Neutrino Program to ν_e appearance

the particular model within which such growth can be realized), the explanation of the MiniBooNE anomaly becomes more consistent. The idea is that one can saturate limits from solar and reactor experiments at MeV scale and exploit the increase of sterile neutrino mixing angles between MeV and GeV scale. Then, the effective mixing angle ($\sin\theta_{\mu e}$) for the transition between muon and electron flavor attains larger values at GeV-scale energies which are relevant for MiniBooNE. This improvement is shown explicitly in Fig. 6.3 in the parameter space of squared sterile neutrino mass $m_s^2 \approx \Delta m^2_{41}$ and $\sin^2 2\theta_{\mu e}$. From the figure, one can infer that the 'RGE' line lies much closer to the regions preferred by MiniBooNE when compared to the line labeled as 'Standard'.

Summary. We have considered renormalization group induced effects that can alter PMNS matrix across the scales relevant for neutrino production and detection. We have discussed phenomenological consequences in the standard three-neutrino picture at T2K, NOvA and IceCube and have also demonstrated that the eV-scale sterile neutrino explanation of the MiniBooNE anomaly becomes more consistent if the mixing angles change with energy.

References

1. B. Dasguptam, J. Kopp, Sterile neutrinos. Phys. Rept. **928**, 1–63 (2021). https://doi.org/10.1016/j.physrep.2021.06.002. arXiv:2106.05913 [hep-ph]
2. Y. Farzan, M. Tortola, Neutrino oscillations and non-standard interactions. Front. Phys. **6**, 10 (2018). https://doi.org/10.3389/fphy.2018.00010. arXiv:1710.09360 [hep-ph]

3. K.S. Babu, V. Brdar, A. de Gouvêa, P.A.N. Machado, Energy-dependent neutrino mixing parameters at oscillation experiments. Phys. Rev. D **105**(11), 115014 (2022). https://doi.org/10.1103/PhysRevD.105.115014. arXiv:2108.11961 [hep-ph]
4. I. Esteban, M.C. Gonzalez-Garcia, M. Maltoni, T. Schwetz, A. Zhou, The fate of hints: updated global analysis of three-flavor neutrino oscillations. JHEP **09**, 178 (2020). https://doi.org/10.1007/JHEP09(2020)178. arXiv:2007.14792 [hep-ph]
5. R. Abbasi et al., [IceCube], The IceCube high-energy starting event sample: description and flux characterization with 7.5 years of data. Phys. Rev. D **104**, 022002 (2021). https://doi.org/10.1103/PhysRevD.104.022002. arXiv:2011.03545 [astro-ph.HE]
6. A.A. Aguilar-Arevalo et al., [MiniBooNE], Updated MiniBooNE neutrino oscillation results with increased data and new background studies. Phys. Rev. D **103**(5), 052002 (2021). https://doi.org/10.1103/PhysRevD.103.052002. arXiv:2006.16883 [hep-ex]
7. V. Brdar, J. Kopp, Can standard model and experimental uncertainties resolve the MiniBooNE anomaly?. Phys. Rev. D **105**(11), 115024 (2022). https://doi.org/10.1103/PhysRevD.105.115024. arXiv:2109.08157 [hep-ph]
8. K.J. Kelly, J. Kopp, More ingredients for an altarelli cocktail at MiniBooNE. arXiv:2210.08021 [hep-ph]
9. M. Dentler, Á. Hernández-Cabezudo, J. Kopp, P.A.N. Machado, M. Maltoni, I. Martinez-Soler, T. Schwetz, Updated global analysis of neutrino oscillations in the presence of eV-scale sterile neutrinos. JHEP **08**, 010 (2018). https://doi.org/10.1007/JHEP08(2018)010. arXiv:1803.10661 [hep-ph]]
10. K.S. Babu, V. Brdar, A. de Gouvêa, P.A.N. Machado, Addressing the short-baseline neutrino anomalies with energy-dependent mixing parameters. arXiv:2209.00031 [hep-ph]

Chapter 7
Theory of Inclusive B Decays

Gil Paz

Abstract In this talk I briefly review the current status of the theory of the inclusive B decays: $\bar{B} \to X_c\, \ell\, \bar{\nu}_\ell$, $\bar{B} \to X_u\, \ell\, \bar{\nu}_\ell$, and $\bar{B} \to X_s\, \gamma$. I try to answer three questions: what is the current "state of the art", can the theoretical prediction be improved, and will it lead to a smaller theoretical uncertainty.

7.1 Introduction

Why study inclusive B decays? First, they are part of flavor physics which allows access to new physics at scales beyond reach of current colliders. For example, $K - \bar{K}$ mixing and $B - \bar{B}$ mixing probe scales above hundreds of TeV [1, 2]. Second, there is a persistent tension between the extraction of $|V_{cb}|$ and $|V_{ub}|$ from inclusive B decays on the one hand and the extraction of $|V_{cb}|$ and $|V_{ub}|$ from exclusive B decays on the other. The inclusive values are consistently larger than the exclusive ones, see every biennial edition of the Review of Particle Physics from 2006 [3] to 2022 [4]. Third, they test basic QFT tools such as factorization theorems and the Operator Product Expansion (OPE) to higher orders both perturbatively and non-perturbatively. For example, the OPE for $\bar{B} \to X_c\, \ell\, \bar{\nu}_\ell$ is now known perturbative to third order and non-perturbative to fourth order, see for example the talks of Keri Vos and Matteo Fael at [5]. Fourth, they are a "window" to non-perturbative physics. For example, at leading twist the $\bar{B} \to X_s\, \gamma$ photon spectrum *is* the B-meson b-quark pdf.

How do we make theoretical predictions? At energies below m_W, m_Z and m_t the effective Hamiltonian is known, see [6]. For example, for $\bar{B} \to X_s\, \gamma$ it is

G. Paz (✉)
Department of Physics and Astronomy, Wayne State University, Detroit, Michigan 48201, USA
e-mail: gilpaz@wayne.edu

G. Ricciardi et al. (eds.), *8th Workshop on Theory, Phenomenology and Experiments in Flavour Physics*, Springer Proceedings in Physics 292,
https://doi.org/10.1007/978-3-031-30459-0_7

$$\mathcal{H}_{\text{eff}} = \frac{G_F}{\sqrt{2}} \sum_{q=u,c} V_{qb} V_{qs}^* \left(C_1 Q_1^q + C_2 Q_2^q + \sum_{i=3}^{10} C_i Q_i + C_{7\gamma} Q_{7\gamma} + C_{8g} Q_{8g} \right) + \text{h.c.}. \tag{7.1}$$

The coefficients C_i are calculable in perturbation theory while Q_i are operators with non-perturbative matrix elements. In particular,

$$\begin{aligned} Q_1^q &= (\bar{q}b)_{V-A}(\bar{s}q)_{V-A} \quad (q = u, c) \\ Q_{7\gamma} &= \frac{-e}{8\pi^2} m_b \bar{s} \sigma_{\mu\nu}(1+\gamma_5) F^{\mu\nu} b \\ Q_{8g} &= \frac{-g_s}{8\pi^2} m_b \bar{s} \sigma_{\mu\nu}(1+\gamma_5) G^{\mu\nu} b. \end{aligned} \tag{7.2}$$

The main problem is that we know the operators but usually we cannot calculate the matrix elements. This is because the operators contain strongly interacting quarks and gluons. These operators can be *local*, e.g., $\bar{q}(0)\cdots q(0)$, or non-local, e.g., $\bar{q}(0)\cdots q(tn)$, where n is a light-like vector. In particular, for inclusive B decays we encounter "diagonal" matrix elements between B-meson states. The operator can be local, as in $\langle \bar{B}|\bar{b}\,\mathbf{D}^2\,b|\bar{B}\rangle = 2M_B\mu_\pi^2$, or non local as in $\int_{-\infty}^{\infty} dt\, e^{i\omega t}\langle \bar{B}(v)|\bar{h}(0)\,[0, tn]\,h(tn)|\bar{B}(v)\rangle = (2\pi)(2M_B)S(\omega)$, where h is a heavy quark and $[0, tn]$ a Wilson line.

What to do with the non-perturbative objects? There are several possible strategies. First, if possible, we can extract them *carefully* from data. For example, the parameter μ_π^2 is extracted from $\bar{B} \to X_c\,\ell\,\bar{\nu}_\ell$. Second, sometimes we can calculate them using some non-perturbative method, e.g. Lattice QCD. For example, the B-meson decay constant is calculated this way. Third, we can use symmetries to reduce the number of non-perturbative objects. For example, use $SU(3)$ flavor to calculate exclusive hadronic decays of B mesons. Fourth, when all else fails, use conservative modeling. For example, this is how the non-perturbative error for $\bar{B} \to X_s\,\gamma$ is determined.

An important feature of inclusive B decays is that we have two expansion parameters we can use. A perturbative parameter: α_s which at the scale m_b is about 0.2 and a non-perturbative parameter: Λ_{QCD}/m_b which is about 0.1. Thus observables can be calculated as a double series in these parameters.

How well can we calculate? Answering this question is the goal of this talk. In the following I will try to answer three questions. What is the current "state of the art"? Can the theoretical prediction be improved? Will it lead to a smaller theoretical uncertainty? I will consider three topics: $|V_{cb}|$ and $\bar{B} \to X_c\,\ell\,\bar{\nu}_\ell$ in Sect. 7.2, $|V_{ub}|$ and $\bar{B} \to X_u\,\ell\,\bar{\nu}_\ell$ in Sect. 7.3, and $\bar{B} \to X_s\,\gamma$ in in Sect. 7.4. For more details on the current status, see also the talks at the "Challenges in Semileptonic B Decays" Workshop held in April 2022 in Barolo, Italy [5].

7.2 $|V_{cb}|$ and $\bar{B} \to X_c \ell \bar{\nu}_\ell$

Inclusive semi-leptonic $b \to c$ decays, namely, $\bar{B} \to X_c \ell \bar{\nu}_\ell$, are governed by the Hamiltonian,

$$\mathcal{H}_{\text{eff}} = \frac{G_F}{\sqrt{2}} C_1(\mu) V_{cb} \bar{\ell}\gamma_\mu(1-\gamma^5)\nu_\ell \, \bar{c}\gamma^\mu(1-\gamma^5)b. \tag{7.3}$$

Using the optical theorem we can calculate $\bar{B} \to X_c \ell \bar{\nu}_\ell$ as an OPE. Up to order $1/m_b^2$ we have symbolically $\Gamma \sim c_0\langle O_0\rangle + c_2^j\langle O_2^j\rangle/m_b^2 + \cdots$, with summation over j implied. In this expression $c_0\langle O_0\rangle$ is a free quark decay. At tree level it is the same as $\mu \to e\,\bar{\nu}_e \nu_\mu$. The coefficients c_i^j are perturbative and can be calculated as a series in α_s. The matrix elements $\langle O_i^j\rangle$ are non perturbative and can be extracted from experiment. For example, $\langle O_0\rangle = \langle\bar{B}|\bar{b}b|\bar{B}\rangle = 2M_B$, $\langle O_2^{\text{kin.}}\rangle = \langle\bar{B}|\bar{b}(iD)^2 b|\bar{B}\rangle$ is related to the HQET parameter μ_π^2, and $\langle O_2^{\text{mag.}}\rangle = \langle\bar{B}|\bar{b}\,\sigma_{\mu\nu}G^{\mu\nu}b|\bar{B}\rangle$ is related to the HQET parameter μ_G^2 and can be extracted from the $B^* - B$ mass difference.

The current "state of the art" is that corrections up to and including $1/m_b^5$ are known. Considering the Wilson coefficients at tree level we have: at $1/m_b^0$ one operator, at $1/m_b$ zero operators, at $1/m_b^2$ two operators [7, 8], at order $1/m_b^3$ two operators [9], at $1/m_b^4$: nine operators, and at $1/m_b^5$: eighteen operators [10]. Are these all the possible operators at each order in $1/m_b$?

This question was answered in [11], where a method was presented that can list such bilinear operators, in principle, to *arbitrary* dimension. More generally it gives NRQED and NRQCD bilinear operators, to arbitrary dimension. See also the work of [12] using Hilbert series. It turns out that the counting above does not include all the possible operators. Allowing for Wilson coefficients at order α_s or higher, at $1/m_b^4$ there are eleven operators instead of nine, and at $1/m_b^5$ there are twenty five operators instead of eighteen. These are unknown but are extremely small. For example, a $1/m_b^4$ operator with an α_s coefficient would give a relative contribution of $\alpha_s\left(\Lambda_{\text{QCD}}/m_b\right)^4 \sim 0.2\cdot(0.1)^4 \sim 10^{-5}$. The $1/m_b^4$, $1/m_b^5$ matrix elements with tree level Wilson coefficients were extracted from $\bar{B} \to X_c\ell\bar{\nu}_\ell$ in [49]. The authors of [49] find that "The higher power corrections have a minor effect on $|V_{cb}|$ ... There is a -0.25% reduction in $|V_{cb}|$".

What is the current "state of the art"? As of 2021

$$\Gamma \sim c_0\langle O_0\rangle + c_2^j\frac{\langle O_2^j\rangle}{m_b^2} + c_3^j\frac{\langle O_3^j\rangle}{m_b^3} + c_4^j\frac{\langle O_4^j\rangle}{m_b^4} + c_5^j\frac{\langle O_5^j\rangle}{m_b^5} + \cdots, \tag{7.4}$$

where c_0 is known at $\mathcal{O}(\alpha_s^0)$, at $\mathcal{O}(\alpha_s^1)$ [13, 14], at $\mathcal{O}(\alpha_s^2)$ [15, 16], and at $\mathcal{O}(\alpha_s^3)$ [17] for the total rate, c_2^j are known at $\mathcal{O}(\alpha_s^0)$ [7, 8] and at $\mathcal{O}(\alpha_s^1)$ [18–21], c_3^j known at $\mathcal{O}(\alpha_s^0)$ [9], and at $\mathcal{O}(\alpha_s^1)$ [22, 23] for selected observables, and c_4^j and c_5^j are known at $\mathcal{O}(\alpha_s^0)$ [10]. The state of the art value of inclusive $|V_{cb}|$ is $42.16(51)\cdot 10^{-3}$ [24], while the 2022 Review of Particle Physics lists the value of exclusive $|V_{cb}|$ as $39.4(8)\cdot 10^{-3}$ [4]. The exclusive/inclusive $|V_{cb}|$ puzzle remains.

Can the theoretical prediction be improved? Yes, for example c_3^j at $\mathcal{O}(\alpha_s^1)$ will probably be calculated in the coming years for the fully differential decay rate. Will it lead to a smaller theoretical uncertainty? Probably. There could be other improvements. For example, the use of the leptonic invariant mass spectrum that depends on less parameters [25]. This method was very recently implemented in [26] finding $|V_{cb}| = 41.69(63) \cdot 10^{-3}$.

7.3 $|V_{ub}|$ and $\bar{B} \to X_u \ell \bar{\nu}_\ell$

$|V_{ub}|$ plays an important role in the unitarity triangle fit. Like $|V_{cb}|$, $|V_{ub}|$ inclusive is larger than $|V_{ub}|$ exclusive. The 2022 Review of Particle Physics lists for inclusive $|V_{ub}|$ the value $(4.13 \pm 0.12^{+0.13}_{-0.14} \pm 0.18) \cdot 10^{-3}$ and for exclusive $|V_{ub}|$ the value $(3.70 \pm 0.10 \pm 0.12) \cdot 10^{-3}$ [4].

If we could measure the total $\Gamma(\bar{B} \to X_u \, l \, \bar{\nu})$ we could use a *local* OPE, just like for $\bar{B} \to X_c \, \ell \, \bar{\nu}_\ell$, see Eq. (7.4). Since $\Gamma(\bar{B} \to X_c \ell \, \bar{\nu}_\ell) \gg (\bar{B} \to X_u \ell \, \bar{\nu}_\ell)$ the total rate *cannot* be measured and one needs to cut the charm background. For example, by requiring that $M_X^2 < M_D^2 \sim m_b \, \Lambda_{\rm QCD}$. This is not inclusive enough for a local OPE, but a non-local OPE is still possible. More generally, one can distinguish three regions depending on the size of M_X^2. The region $M_X^2 \sim m_b^2$ is the "OPE region", where we have a local OPE. The region $M_X^2 \sim m_b \, \Lambda_{\rm QCD}$ is the "end-point region" where we have a non-local OPE. The region $M_X^2 \sim \Lambda_{\rm QCD}^2$ is the "resonance region" where no inclusive description is possible and one must use an exclusive description.

For $\bar{B} \to X_u \, l \, \bar{\nu}$ we therefore have a non-local OPE, where the spectra is described in terms on non-local matrix elements called "shape functions" or "soft functions". Schematically we have

$$d\Gamma \sim H \cdot J \otimes S + \frac{1}{m_b} \sum_i H \cdot J \otimes s_i + \cdots . \tag{7.5}$$

At leading power in $\Lambda_{\rm QCD}/m_b$, $d\Gamma$ factorizes to a product of a "hard" (H) and "jet" (J) functions convoluted with a a non-perturbative "shape function" (S) [28–30]. Intuitively S is the B-meson b-quark pdf. A similar factorization formula holds for $\bar{B} \to X_s \, \gamma$ at the photon energy end-point region. As a result, at leading power in $\Lambda_{\rm QCD}/m_b$, S is the $\bar{B} \to X_s \, \gamma$ photon spectrum. As an example, see its recent extraction by the SIMBA Collaboration [27].

What about $\Lambda_{\rm QCD}/m_b$ corrections? At this order several subleading shape functions (SSF) appear denoted in Eq. (7.5) by s_i [31–35]. They are analogous to "higher twist" contributions for deep inelastic scattering. Different linear combinations of SSF appear for $\bar{B} \to X_u \, \ell \, \bar{\nu}_\ell$ and $\bar{B} \to X_s \, \gamma$. Furthermore, $\bar{B} \to X_s \, \gamma$ has unique SSF, called "resolved photon contributions", see Sect. 7.4.

Moments of the shape functions are related to "universal" matrix elements that appear also for $\bar{B} \to X_c \, \ell \, \bar{\nu}_\ell$. For example, the first moment of S is related to the b-quark mass and the second moment is related to μ_π^2.

There are different theoretical frameworks for $|V_{ub}|$ extractions. They use similar perturbative inputs, currently at $\mathcal{O}(\alpha_s)$, but they differ in how they extract (or model) S and how they treat power corrections. Current theoretical frameworks used by experiments are designated by acronyms. Thus we have BLNP [36], DGE [37], GGOU [38], and ADFR [39]. A recent extraction of inclusive $|V_{ub}|$ from Belle data finds almost no difference between the $|V_{ub}|$ values from these frameworks [40]. See also Francesco Tenchini's talk at this workshop. We should keep in mind though that these frameworks were developed before 2010.

Can the theoretical prediction be improved? Yes, many NNLO calculations are known. The free quark differential decay rate was calculated at $\mathcal{O}(\alpha_s^2)$ in [41]. The hard function H is known at $\mathcal{O}(\alpha_s^2)$ [42–45]. The jet function J was calculated at $\mathcal{O}(\alpha_s^2)$ in [46] and at $\mathcal{O}(\alpha_s^3)$ in [47]. Corrections that scale like $\mathcal{O}(\alpha_s) \times \mathcal{O}(\Lambda_{\rm QCD}/m_b)$ in the form of subleading jet function (j_i) were calculated in [48]. Moments of the leading (S) and subleading shape functions (s_i) can be calculated using the data from [49] and the method of [11] (see also the appendix of [50]). For frameworks that use $\bar{B} \to X_s \, \gamma$, there is also the calculation of the resolved photon contributions in [51]. All of these are not fully combined yet. Will it lead to a smaller theoretical uncertainty? Not necessarily.

7.4 $\bar{B} \to X_s \, \gamma$

$\bar{B} \to X_s \, \gamma$ is a well known probe of physics beyond the standard model. The 2022 edition of the Review of Particle Physics lists the CP averaged branching ratio as $(3.49 \pm 0.19) \cdot 10^{-4}$ [4]. The standard model prediction in 2015 was $(3.36 \pm 0.23) \cdot 10^{-4}$ [52]. The largest uncertainty of this prediction, about 5%, is non-perturbative from "resolved photons contributions". These are contributions in which the photon does not couple directly to the weak vertex [51]. They are parameterized by fields that are localized on two different light cones. In some sense they are a precursor to the contributions discussed in Matthias Neubert's talk at this workshop.

At $\mathcal{O}(\Lambda_{\rm QCD}/m_b)$ the resolved photon contributions arise from the interference of $Q_1 - Q_{7\gamma}$, $Q_{7\gamma} - Q_{8g}$ and $Q_{8g} - Q_{8g}$. See Eq. (7.2) for their definitions. The standard model CP asymmetry is dominated by $Q_1^q - Q_{7\gamma}$: $-0.6\% < \mathcal{A}_{X_s\gamma}^{\rm SM} < 2.8\%$ [53], while the 2022 edition of the Review of Particle Physics lists the range from experiment as $\mathcal{A}_{X_s\gamma} = 1.5\% \pm 1.1\%$ [4].

Can these be improved? The $Q_{8g} - Q_{8g}$ contribution is hard to improve, but beyond the power suppression it is suppressed by the square of the s-quark charge. The $Q_{7\gamma} - Q_{8g}$ contribution is constrained by isospin asymmetry $\bar{B}^{0/\pm} \to X_s \, \gamma$ [54]. This uncertainty was reduced by a Belle measurement [55]. The $Q_1 - Q_{7\gamma}$ contribution depends on a non-perturbative function $g_{17}(\omega, \omega_1)$ whose moments can be extracted from $\bar{B} \to X_c \, \ell \, \bar{\nu}_\ell$ OPE. The 2010 analysis of [51] only had two non-zero moments. A 2019 analysis [56] by Ayesh Gunawardana and me added six non-zero moments using the data from [49] and the methods of [11]. Using moments we can better model the $Q_1 - Q_{7\gamma}$ resolved photon contribution. The new estimate reduced the uncertainty on the total rate by up to 50%, but increased the uncertainty on the CP asymmetry by about 33%. These improvements were included in the 2020 updated standard model prediction of $(3.40 \pm 0.17) \cdot 10^{-4}$ [57]. Using different models, including some $\mathcal{O}\left(\Lambda_{\rm QCD}^2/m_b^2\right)$ corrections and larger m_c range, a smaller reduction was found in [58].

Can the theoretical prediction be improved? Yes, the m_c dependence can only be really controlled by an NLO analysis of the $Q_1 - Q_{7\gamma}$ contribution. Will it lead to a smaller theoretical uncertainty? Not necessarily.

7.5 Conclusions

Flavor physics probes very high scales and advanced theoretical tools. This decade will be very exciting with, e.g., LHCb and Belle II data. These supply puzzles and tensions that motivate further theoretical work. A big challenge is controlling non-perturbative effects. I discussed the "state of the art" of $|V_{cb}|$ and $\bar{B} \to X_c\, \ell\, \bar{\nu}_\ell$, $|V_{ub}|$ and $\bar{B} \to X_u\, \ell\, \bar{\nu}_\ell$, and $\bar{B} \to X_s\, \gamma$. While the theory is fairly advanced, there is still room for theoretical improvements. These will lead to a more robust theoretical uncertainty but not necessarily to a smaller one.

Acknowledgements I thank the organizers of the 8th Workshop on Theory, Phenomenology and Experiments in Flavour Physics: Neutrinos, Flavor Physics and Beyond (FPCapri22) for the invitation to give this talk. This work was supported in part by the DOE grant DE-SC0007983.

References

1. M. Bona et al., UTfit. JHEP **03**, 049 (2008). arXiv:0707.0636 [hep-ph]
2. A. Bevan, M. Bona, M. Ciuchini, D. Derkach, E. Franco, V. Lubicz, G. Martinelli, F. Parodi, M. Pierini, C. Schiavi, et al. arXiv:1411.7233 [hep-ph]
3. W.M. Yao et al., Particle Data Group. J. Phys. G **33**, 1–1232 (2006)
4. R.L. Workman et al., [Particle Data Group], PTEP **2022**, 083C01 (2022)
5. "Challenges in Semileptonic B Decays" Workshop held in April 2022 in Barolo, Italy. https://indico.cern.ch/event/851900/
6. A.J. Buras, arXiv:hep-ph/9806471 [hep-ph]
7. B. Blok, L. Koyrakh, M.A. Shifman, A.I. Vainshtein, Phys. Rev. D **49**, 3356 (1994) [erratum: Phys. Rev. D **50**, 3572 (1994)]. arXiv:hep-ph/9307247 [hep-ph]
8. A.V. Manohar, M.B. Wise, Phys. Rev. D **49**, 1310–1329 (1994). arXiv:hep-ph/9308246 [hep-ph]
9. M. Gremm, A. Kapustin, Phys. Rev. D **55**, 6924–6932 (1997). arXiv:hep-ph/9603448 [hep-ph]
10. T. Mannel, S. Turczyk, N. Uraltsev, JHEP **11**, 109 (2010). arXiv:1009.4622 [hep-ph]
11. A. Gunawardana, G. Paz, JHEP **07**, 137 (2017). arXiv:1702.08904 [hep-ph]
12. A. Kobach, S. Pal, Phys. Lett. B **772**, 225–231 (2017). arXiv:1704.00008 [hep-ph]

13. M. Trott, Phys. Rev. D **70**, 073003 (2004). arXiv:hep-ph/0402120 [hep-ph]
14. V. Aquila, P. Gambino, G. Ridolfi, N. Uraltsev, Nucl. Phys. B **719**, 77–102 (2005). arXiv:hep-ph/0503083 [hep-ph]
15. K. Melnikov, Phys. Lett. B **666**, 336–339 (2008). arXiv:0803.0951 [hep-ph]
16. A. Pak, A. Czarnecki, Phys. Rev. D **78**, 114015 (2008). arXiv:0808.3509 [hep-ph]
17. M. Fael, K. Schönwald, M. Steinhauser, Phys. Rev. D **104**(1), 016003 (2021). arXiv:2011.13654 [hep-ph]
18. T. Becher, H. Boos, E. Lunghi, JHEP **12**, 062 (2007). arXiv:0708.0855 [hep-ph]
19. A. Alberti, T. Ewerth, P. Gambino, S. Nandi, Nucl. Phys. B **870**, 16–29 (2013). arXiv:1212.5082 [hep-ph]
20. A. Alberti, P. Gambino, S. Nandi, JHEP **01**, 147 (2014). arXiv:1311.7381 [hep-ph]
21. T. Mannel, A.A. Pivovarov, D. Rosenthal, Phys. Rev. D **92**(5), 054025 (2015). arXiv:1506.08167 [hep-ph]
22. T. Mannel, A.A. Pivovarov, Phys. Rev. D **100**(9), 093001 (2019). arXiv:1907.09187 [hep-ph]
23. T. Mannel, D. Moreno. A.A. Pivovarov, Phys. Rev. D **105**(5), 054033 (2022). arXiv:2112.03875 [hep-ph]
24. M. Bordone, B. Capdevila, P. Gambino, Phys. Lett. B **822**, 136679 (2021). arXiv:2107.00604 [hep-ph]
25. M. Fael, T. Mannel, K. Keri Vos, JHEP **02**, 177 (2019). arXiv:1812.07472 [hep-ph]
26. F. Bernlochner, M. Fael, K. Olschewsky, E. Persson, R. van Tonder, K.K. Vos, M. Welsch, JHEP **10**, 068 (2022). arXiv:2205.10274 [hep-ph]
27. F.U. Bernlochner et al. [SIMBA], Phys. Rev. Lett. **127**(10), 102001 (2021). arXiv:2007.04320 [hep-ph]
28. M. Neubert, Phys. Rev. D **49**, 3392 (1994). [hep-ph/9311325]
29. M. Neubert, Phys. Rev. D **49**, 4623 (1994). [hep-ph/9312311]
30. I.I.Y. Bigi, M.A. Shifman, N.G. Uraltsev, A.I. Vainshtein, Int. J. Mod. Phys. A **9**, 2467 (1994). [hep-ph/9312359]
31. K.S.M. Lee, I.W. Stewart, Nucl. Phys. B **721**, 325 (2005). [hep-ph/0409045]
32. S.W. Bosch, M. Neubert, G. Paz, JHEP **0411**, 073 (2004). [hep-ph/0409115]
33. M. Beneke, F. Campanario, T. Mannel, B.D. Pecjak, JHEP **0506**, 071 (2005). [hep-ph/0411395]
34. C.W. Bauer, M.E. Luke, T. Mannel, Phys. Rev. D **68**, 094001 (2003). [hep-ph/0102089]
35. A.K. Leibovich, Z. Ligeti, M.B. Wise, Phys. Lett. B **539**, 242 (2002). [hep-ph/0205148]
36. B.O. Lange, M. Neubert, G. Paz, Phys. Rev. D **72**, 073006 (2005). ([arXiv:hep-ph/0504071 [hep-ph]])
37. J.R. Andersen, E. Gardi, JHEP **01**, 097 (2006). ([arXiv:hep-ph/0509360 [hep-ph]])
38. P. Gambino, P. Giordano, G. Ossola, N. Uraltsev, JHEP **10**, 058 (2007). ([arXiv:0707.2493 [hep-ph]])
39. U. Aglietti, F. Di Lodovico, G. Ferrera, G. Ricciardi, Eur. Phys. J. C **59**, 831–840 (2009). ([arXiv:0711.0860 [hep-ph]])
40. L. Cao *et al.* [Belle], Phys. Rev. D **104**, no.1, 012008 (2021) [arXiv:2102.00020 [hep-ex]]
41. M. Brucherseifer, F. Caola, K. Melnikov, Phys. Lett. B **721**, 107–110 (2013). ([arXiv:1302.0444 [hep-ph]])
42. R. Bonciani, A. Ferroglia, JHEP **0811**, 065 (2008). ([arXiv:0809.4687 [hep-ph]])
43. H.M. Asatrian, C. Greub, B.D. Pecjak, Phys. Rev. D **78**, 114028 (2008). ([arXiv:0810.0987 [hep-ph]])
44. M. Beneke, T. Huber, X.-Q. Li, Nucl. Phys. B **811**, 77 (2009). ([arXiv:0810.1230 [hep-ph]])
45. G. Bell, Nucl. Phys. B **812**, 264 (2009). ([arXiv:0810.5695 [hep-ph]])
46. T. Becher and M. Neubert, Phys. Lett. B **637**, 251-259 (2006) [arXiv:hep-ph/0603140 [hep-ph]]
47. R. Brüser, Z. L. Liu and M. Stahlhofen, Phys. Rev. Lett. **121**, no.7, 072003 (2018) [arXiv:1804.09722 [hep-ph]]
48. G. Paz, JHEP **06**, 083 (2009). ([arXiv:0903.3377 [hep-ph]])
49. P. Gambino, K.J. Healey, S. Turczyk, Phys. Lett. B **763**, 60–65 (2016). ([arXiv:1606.06174 [hep-ph]])
50. J. Heinonen and T. Mannel, [arXiv:1609.01334 [hep-ph]]

51. M. Benzke, S.J. Lee, M. Neubert, G. Paz, JHEP **08**, 099 (2010). ([arXiv:1003.5012 [hep-ph]])
52. M. Misiak, H. M. Asatrian, R. Boughezal, M. Czakon, T. Ewerth, A. Ferroglia, P. Fiedler, P. Gambino, C. Greub and U. Haisch, *et al.* Phys. Rev. Lett. **114**, no.22, 221801 (2015) [arXiv:1503.01789 [hep-ph]]
53. M. Benzke, S.J. Lee, M. Neubert, G. Paz, Phys. Rev. Lett. **106**, 141801 (2011). ([arXiv:1012.3167 [hep-ph]])
54. M. Misiak, Acta Phys. Polon. B **40**, 2987–2996 (2009). ([arXiv:0911.1651 [hep-ph]])
55. S. Watanuki *et al.* [Belle], Phys. Rev. D **99**, no.3, 032012 (2019) [arXiv:1807.04236 [hep-ex]]
56. A. Gunawardana, G. Paz, JHEP **11**, 141 (2019). ([arXiv:1908.02812 [hep-ph]])
57. M. Misiak, A. Rehman, M. Steinhauser, JHEP **06**, 175 (2020). ([arXiv:2002.01548 [hep-ph]])
58. M. Benzke, T. Hurth, Phys. Rev. D **102**, 114024 (2020). ([arXiv:2006.00624 [hep-ph]])

Chapter 8
Heavy Neutral Leptons in the N_RSMEFT and the High-Luminosity LHC

Arsenii Titov

Abstract We study heavy neutral leptons (HNLs) in the N_RSMEFT, the effective field theory of the Standard Model extended with gauge singlet right-handed fermions. Four-fermion operators which contain two HNLs and two quarks can lead to a sizeable enhancement of the HNL pair-production cross section, compared to the minimal case where HNLs are produced only via their mixing with the active neutrinos. We calculate the expected sensitivities for the ATLAS detector and the future far-detector experiments at the high-luminosity LHC: AL3X, ANUBIS, CODEX-b, FASER, MATHUSLA and MoEDAL-MAPP. We find that new physics scales up to 22 TeV and active-heavy mixing squared as small as 10^{-24} could be probed by some of these experiments. Next, we consider four-fermion operators with a single HNL. Such operators might dominate both HNL production and decay, and we find that new physics scales in excess of 20 TeV could be tested at ATLAS.

8.1 Introduction

Heavy neutral leptons (HNLs) or right-handed (RH) neutrinos, N_R, are present in many extensions of the Standard Model (SM) able to account for non-zero masses of light neutrinos. In the canonical type I seesaw mechanism [1–5], these new fermions are very heavy, with masses of 10^{13}–10^{15} GeV. However, in other seesaw variants, such as inverse [6, 7] or linear [8, 9] seesaw, their masses can be below or around the electroweak (EW) scale $v = 246$ GeV. In the minimal, renormalisable set-up, HNLs interact with the SM via their mixing with the active neutrinos. However, viewing the SM as an effective field theory (EFT) (see Ref. [10] for a review), HNLs can have

A. Titov (✉)
Departament de Física Teòrica, Universitat de València, IFIC, Universitat de València–CSIC, Dr. Moliner 50, E–46100 Burjassot, Spain
e-mail: arsenii.titov@ific.uv.es

G. Ricciardi et al. (eds.), *8th Workshop on Theory, Phenomenology and Experiments in Flavour Physics*, Springer Proceedings in Physics 292,
https://doi.org/10.1007/978-3-031-30459-0_8

new non-renormalisable interactions with the SM fields, respecting the SM gauge symmetry $SU(3)_C \times SU(2)_L \times U(1)_Y$. The EFT of the SM extended with HNLs is known as the N_RSMEFT [11–13].[1]

For certain values of the HNL mass, m_N, and active-heavy mixing, $V_{\alpha N}$, $\alpha = e, \mu, \tau$, or a non-renormalisable coupling responsible for the HNL decay, HNLs can be long-lived. In this case, they could lead to variety of exotic signatures as their decay products are displaced from their production position [14]. These signatures can be searched for at the LHC local detectors, in particular, at ATLAS and CMS. In addition, in the past few years a number of far detectors specifically designed to search for long-lived particles (LLPs) have been proposed: AL3X [15], ANUBIS [16], CODEX-b [17], FASER and FASER2 [18, 19], MoEDAL-MAPP1 and MAPP2 [20, 21], and MATHUSLA [22–24]. The reach of these future experiments to the minimal HNLs has been studied in a number of works, see e.g. Refs. [25–29].

In this contribution, based on Refs. [30, 31], we first give a brief introduction to the N_RSMEFT and then describe the reach of the local and far detectors at the high-luminosity LHC (HL-LHC) to the HNLs in this EFT.

8.2 The N_RSMEFT

If HNLs with masses below or around the EW scale exist in nature, the effects of new multi-TeV physics at much lower energies can be systematically described in terms of an EFT built out of the SM fields and N_R. At the renormalisable level, in addition to the SM operators, there are a Majorana mass term for N_R and a mass-dimension $d = 4$ operator describing the fermion portal:

$$\mathcal{L}_{\text{ren}} = \mathcal{L}_{\text{SM}} + \overline{N_R} i \partial\!\!\!/ N_R - \left[\frac{1}{2} \overline{N_R^c} M_N N_R + \overline{L} \tilde{H} Y_N N_R + \text{h.c.} \right], \tag{8.1}$$

where L stands for the SM lepton doublets, H is the Higgs doublet ($\tilde{H} = \epsilon H^*$, ϵ is the totally antisymmetric tensor), and $N_R^c \equiv C \overline{N_R}^T$, with C being the Dirac charge conjugation matrix. The Majorana mass matrix M_N is a symmetric $n_N \times n_N$ matrix, with n_N denoting the number of HNL generations, and Y_N is a generic $3 \times n_N$ matrix of Yukawa couplings.

Upon including non-renormalisable interactions $\mathcal{O}_i^{(d)}$ with $d \geq 5$, the full Lagrangian reads

$$\mathcal{L} = \mathcal{L}_{\text{ren}} + \sum_{d \geq 5} \frac{1}{\Lambda^{d-4}} \sum_i c_i^{(d)} \mathcal{O}_i^{(d)}, \tag{8.2}$$

[1] Other names, as NSMEFT, SMNEFT or νSMEFT, are also used in the literature.

Table 8.1 LNC four-fermion quark-N_R operators in the N_RSMEFT, containing either two (upper part) or one (lower part) N_R

Name	Structure
$\mathcal{O}_{dN}$	$(\overline{d_R}\gamma^\mu d_R)(\overline{N_R}\gamma_\mu N_R)$
$\mathcal{O}_{uN}$	$(\overline{u_R}\gamma^\mu u_R)(\overline{N_R}\gamma_\mu N_R)$
$\mathcal{O}_{QN}$	$(\overline{Q}\gamma^\mu Q)(\overline{N_R}\gamma_\mu N_R)$
Name	Structure (+ h.c.)
$\mathcal{O}_{duNe}$	$(\overline{d_R}\gamma^\mu u_R)(\overline{N_R}\gamma_\mu e_R)$
$\mathcal{O}_{LNQd}$	$(\overline{L}N_R)\epsilon(\overline{Q}d_R)$
$\mathcal{O}_{LdQN}$	$(\overline{L}d_R)\epsilon(\overline{Q}N_R)$
$\mathcal{O}_{QuNL}$	$(\overline{Q}u_R)(\overline{N_R}L)$

where $c_i^{(d)}$ are the Wilson coefficients, and the second sum goes over all independent interactions of a given dimension d. At $d = 5$, in addition to the renowned Weinberg operator composed of L and H [32], one finds two more operators that involve N_R [11, 12].

At $d = 6$, in addition to the pure SMEFT operators [33], there are five operators involving two fermions (at least one of which is N_R) and bosons, eleven baryon and lepton-number-conserving (LNC) four-fermion interactions, one lepton-number-violating (LNV) operator, and two operators that violate both baryon and lepton number [13].[2] We are interested in the effects of the LNC four-fermion quark-N_R operators. We list them in Table 8.1.

8.3 Experiments and Numerical Procedure

The pair-N_R operators shown in the upper part of Table 8.1 may enhance pair-production of HNLs at the LHC, while for one generation of HNLs, they cannot mediate the HNL decay by themselves. However, HNLs can decay via active-heavy mixing, $V_{\alpha N}$. As we will see, such a set-up can be probed at the HL-LHC.

For the ATLAS experiment, our analysis focuses on the HNLs decaying to an electron and two jets, $N \to ejj$. The total number of signal events at ATLAS with the luminosity $\mathcal{L}$ is calculated as

$$N_{\text{sig.}}^{\text{ATLAS}} = \sigma \cdot \mathcal{L} \cdot \text{Br}\,(N \to ejj) \cdot 2 \cdot \epsilon\,, \tag{8.3}$$

where ϵ labels the efficiency of event selection including the ATLAS detector geometry. This efficiency depends on both the HNL mass and proper lifetime. The production cross section, σ, is a function of m_N and the Wilson coefficients of the

[2] Here, we count the operator types, i.e. we do not take into account the flavour structure and do not count hermitian conjugates.

operators switched on, and the branching ratio $\text{Br}(N \to ejj)$ depends on m_N only. The efficiency includes detector acceptance and corresponds to the efficiency for reconstructing one displaced vertex in an event after all the selection criteria have been applied, for details see Sect. 3.2 of Ref. [30]. Note that the factor of 2 arises from the fact that we simply require either of the two HNLs in each event to decay to ejj with a displacement, while disregarding how the other HNL decays.

Under the zero background assumption, 95% C.L. exclusion limits in the $|V_{eN}|^2$ versus m_N plane can be derived by requiring three signal events. These limits are provided and discussed in Sect. 8.4. We note that constraints on mixing in the muon sector can also be obtained if the HNL decays to μjj. In this case, we expect similar exclusion reach. For mixing in the tau sector, we expect the sensitivity reach to be less powerful as a result of worse efficiencies and experimental difficulties in reconstructing (displaced) tau leptons.

In recent years, a series of far-detector experiments have been proposed with a distance of about 5–500 m from the various interaction points (IPs) at the LHC: AL3X, ANUBIS, CODEX-b, FASER and FASER2, MoEDAL-MAPP1 and MAPP2, and MATHUSLA. In particular, MoEDAL-MAPP1 and FASER have been approved and will be taking data during Run 3 of the LHC, whereas the rest of the programme is planned for the HL phase. These experiments are supposed to be shielded from the associated IP by rock, lead, or other material, removing the SM background events effectively. The other background sources include cosmic rays (especially for the on-the-ground experiment MATHUSLA), which can be eliminated by directional cuts. Consequently, in almost all the cases, essentially zero background can be assumed, and correspondingly the 95% C.L. exclusion limits can be derived by requiring three signal events. It is also worth mentioning that these experiments are proposed to be operated in different phases of the LHC and IPs, and so the projected integrated luminosities vary. For instance, MAPP1 is to receive 30 fb^{-1} of integrated luminosity, while FASER2 about 3 ab^{-1}. For a detailed description of the far-detector experiments, we refer the reader to Ref. [29] and the references therein.

The number of signal events at a far detector is computed as follows:

$$N^{\text{FD}}_{\text{sig.}} = 2 \cdot \sigma \cdot \mathcal{L} \cdot \langle P[N \text{ decay in f.v.}] \rangle \cdot \text{Br}\,(N \to \text{vis.})\,, \tag{8.4}$$

where $\langle P[N \text{ decay in f.v.}] \rangle$ denotes the average decay probability of all the simulated HNLs (2 in each event) inside the fiducial volume of a far detector, and $\text{Br}\,(N \to \text{vis.})$ labels the HNL decay branching ratio into visible states, meaning all or some of the decay products are electrically charged (here we only consider the tri-neutrino final state to be invisible). For the computation of $\langle P[N \text{ decay in f.v.}] \rangle$, see Ref. [29]. Concerning decays of HNLs to SM particles via mixing, these have been calculated several times in the literature. In our numerical analysis, we use the decay formulae from Refs. [34, 35].

The single-N_R operators shown in the lower part of Table 8.1 can lead to enhanced HNL production cross section at the LHC, but they also trigger the decay of N to a lepton and two quarks. The total HNL decay width depends on the operator.

Neglecting the masses of the lepton and light quarks, the partial decay width to a charged lepton plus quarks is given by

$$\Gamma(N \to \ell q q') = \frac{c_{\mathcal{O}}^2 m_N^5}{f_{\mathcal{O}} 512 \pi^3 \Lambda^4}, \tag{8.5}$$

with $c_{\mathcal{O}}$ being the Wilson coefficient of the operator $\mathcal{O}$, and $f_{\mathcal{O}}$ the numerical factor depending on the operator type. For $\mathcal{O}_{duNe}$, $f_{\mathcal{O}} = 1$, whereas for $\mathcal{O}_{LNQd}$, $\mathcal{O}_{LdQN}$, and $\mathcal{O}_{QuNL}$, $f_{\mathcal{O}} = 4$. To arrive at the total decay width, one has to add also the final state with neutrinos for all operators, except $\mathcal{O}_{duNe}$. Since the partial width to neutrinos follows the same equation as for charged leptons, this results in total decay widths being twice the partial decay widths given in Eq. (8.5) (again, except for $\mathcal{O}_{duNe}$). Finally, Eq. (8.5) applies to Dirac HNLs. For Majorana HNLs, one has to add also the charged conjugate channels, leading to another factor of 2 for the widths.

The signal topology in the case of the single-N_R operators contains a prompt lepton and a displaced vertex stemming from the N decay to a lepton and quarks. Such a signature can be looked for at the ATLAS detector, specifically in its inner tracker, as it has the capability to reconstruct vertices displaced from the IP by few millimetres to tens of centimetres. The analysis strategy builds up on an earlier work [36]. We refer the reader to Sect. 3 of Ref. [31] for details.

8.4 Results

In this section, we first present and discuss some selected results for the four-fermion pair-N_R operators and then for the single-N_R operators. A detailed discussion of the phenomenology associated to these sets of operators can be found in Refs. [30] and [31], respectively.

8.4.1 Pair-N_R Operators

In Fig. 8.1, we show the estimated 95% C.L. exclusion limits at the 14 TeV LHC in the plane $|V_{eN}|^2$ versus m_N. In the left panel, we focus on one single operator $\mathcal{O}_{dN}$, choosing only the coupling with down quarks, c_{dN}^{11}, to be non-zero. In the right panel, we switch on all the three types of operators, $\mathcal{O}_{dN}$, $\mathcal{O}_{uN}$ and $\mathcal{O}_{QN}$, taking the same value for all the couplings to the first-generation quarks, $c_{dN}^{11} = c_{uN}^{11} = c_{QN}^{11}$. These two choices lead to respectively the smallest and largest possible cross sections for the HNL pair-production at the LHC, and thus, cover the most conservative and the most optimistic scenarios. For the operator coefficients $c_{\mathcal{O}}/\Lambda^2 = 1/(2\text{ TeV})^2$, the experiment MATHUSLA, for an integrated luminosity of 3 ab^{-1}, can reach values of the mixing down to $|V_{eN}|^2 \sim 2 \times 10^{-23}$ for $\mathcal{O}_{dN}$ and $|V_{eN}|^2 \sim 3 \times 10^{-24}$ for the most optimistic scenario in which $\mathcal{O}_{dN}$, $\mathcal{O}_{uN}$ and $\mathcal{O}_{QN}$ are simultaneously

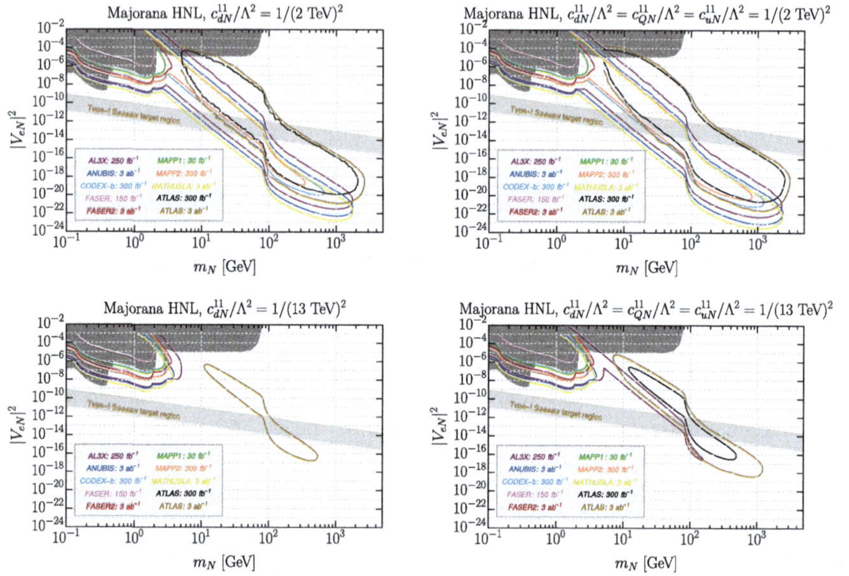

Fig. 8.1 Sensitivity reach of the HL-LHC on $|V_{eN}|^2$ as a function of m_N. The left panel contains results for one single operator $\mathcal{O}_{dN}$. The plots on the right are for the case where all the three pair-N_R effective operators are switched on simultaneously. The *dark grey region* represents the current constraints, and the *light grey band* corresponds to $m_\nu = 10^{-3}$–10^{-1} eV, assuming the type I seesaw relation. Figure from Ref. [30]

switched on. The MATHUSLA experimental sensitivity also extends considerably for the HNL mass reaching up to $m_N \sim 1.8$ (2.3) TeV for the most conservative (optimistic) choice of effective operators. Such a large mass reach is essentially allowed because the HNLs are produced from direct parton collisions of the centre-of-mass energy of 14 TeV. Similar limits can be set by ANUBIS for the same integrated luminosity and around one order of magnitude smaller by AL3X for an integrated luminosity of 250 fb^{-1}. ATLAS, on the other hand, can reach values of mixing down to $|V_{eN}|^2 \sim 10^{-21}$ (2.6×10^{-22}) and masses up to $m_N \sim 2.6$ (3) TeV, for an integrated luminosity of 3 ab^{-1}, for the most conservative (optimistic) scenario. Its lower mass reach is at about 5 GeV, which is due to the $m_{\text{DV}} > 5$ GeV event selection we apply, m_{DV} being the invariant mass of a displaced vertex. This cut is necessary to remove the SM background coming from B-mesons. ATLAS has a sensitivity on $|V_{eN}|^2$ worse than MATHUSLA by almost two orders of magnitude, but it can reach larger m_N. This is mainly due to the distance at which MATHUSLA is planned to be located (~100 m from the IP), which makes it sensitive to higher values of the lifetime, and in turn to smaller mixing $|V_{eN}|^2$ and masses m_N. As we can see, with the inclusion of the effective operators, these experiments can reach limits on the mixing $|V_{eN}|^2$ several orders of magnitude better than the current experimental bounds, represented here by the dark grey region.

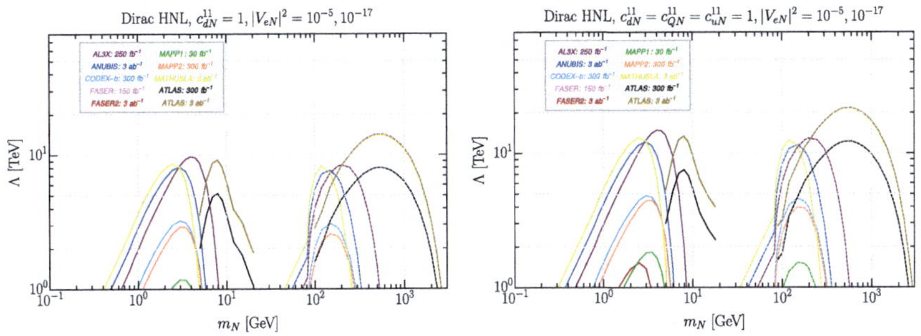

Fig. 8.2 Sensitivity reach of the HL-LHC on the new physics scale Λ as a function of m_N for the pair-N_R operators with the first-generation quarks. The values of active-heavy mixing have been fixed to $|V_{eN}|^2 = 10^{-5}$ (*solid lines*) and 10^{-17} (*dashed lines*). Figure from Ref. [30]

Overall, the shape of the contours shown in Fig. 8.1 can be qualitatively understood as the final efficiency being bounded by the cases when the HNLs are decaying either too promptly (the upper right corner) or too far away (the lower left corner), both outside the detectors' acceptance. In addition, the upper mass reach is limited by the HNL production cross section.

Figure 8.2 depicts the limits in the plane Λ versus m_N for a Dirac HNL, in the cases of $c_{dN}^{11} = 1$ (left panel) and $c_{dN}^{11} = c_{uN}^{11} = c_{QN}^{11} = 1$ (right panel). The values of active-heavy mixing are fixed to be $|V_{eN}|^2 = 10^{-5}$ and 10^{-17}. For values between these extremes, intermediate values of m_N will be tested. This figure shows that the far-detector experiments MATHUSLA, ANUBIS and AL3X may probe the new physics scale up to $\Lambda \sim 8$–15 TeV, depending on the effective operator scenario considered. ATLAS, on the other hand, can probe new physics scale up to $\Lambda \approx 22$ TeV, when we simultaneously switch on the three types of effective operators, and up to $\Lambda \approx 15$ TeV, when we consider the most conservative scenario with one single effective operator $\mathcal{O}_{dN}$. The bounds on Λ in the Majorana HNL case are similar.

8.4.2 Single-N_R Operators

In Fig. 8.3, we show the experimental sensitivity of the ATLAS detector to long-lived HNLs in the Λ versus m_N plane. In our analysis, we have considered the contributions of one operator at a time, setting the value of the corresponding operator coefficient $c_{\mathcal{O}} = 1$, and the rest of the operator coefficients to zero. Here we consider only operators with quarks of the first generation. Note that the numbers in the superscript of e.g. c_{duNe}^{1113} refer to the first-generation quarks (d and u), the lightest N and the tau lepton. As can be seen from this figure, for an integrated luminosity of 3 ab^{-1}, ATLAS can reach values of the new physics scale up to (and above) $\Lambda \approx 20$ TeV for masses $m_N \lesssim 50$ GeV in the case of operators with an electron. The results for

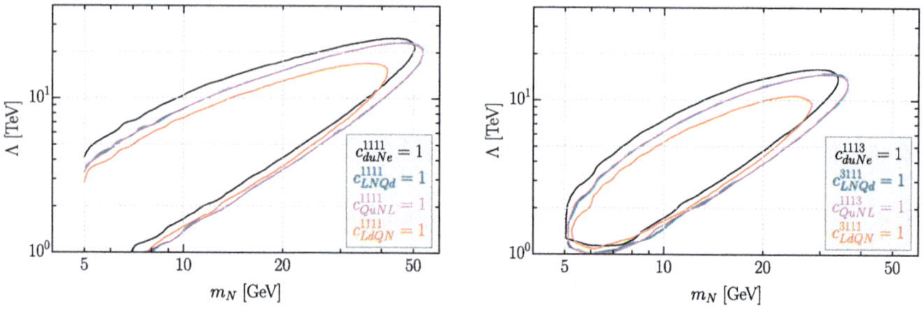

Fig. 8.3 Projected ATLAS exclusion limits on the new physics scale Λ as a function of m_N for the single-N_R operators with the first-generation quarks, assuming an integrated luminosity of 3 ab^{-1}. The plot on the left (right) is for the operators with an electron (a tau lepton). Figure from Ref. [31]

the operators with a muon are very similar, and we do not show them here. In the case of operators with a tau lepton, ATLAS can reach $\Lambda \gtrsim 10$ TeV at masses m_N of tens of GeV. It is worth mentioning that our limits start at $m_N \gtrsim 5$ GeV. The reason is the kinematic cut at $m_{\text{DV}} \geq 5$ GeV imposed in the selection criteria. We also note that the projected exclusion limits are rather similar for the four types of single-N_R operators, and in particular for $\mathcal{O}_{LNQd}$ and $\mathcal{O}_{QuNL}$.

8.5 Summary and Conclusions

In this contribution, based on Refs. [30, 31], we have discussed the phenomenology of HNLs as LLPs in the framework of the N_RSMEFT, the EFT of the SM extended with HNLs (or RH neutrinos), N_R.

First, we have investigated the effects of the four-fermion operators with a pair of HNLs and a pair of quarks. We have considered three different types of these operators and have studied scenarios with only one or all of them being present at the same time. While quantitatively the results depend on the assumptions made about type of operator and/or Wilson coefficient present, qualitatively all the pair-N_R operators behave similarly. They lead to production cross sections at the LHC which are not suppressed by the small mixing of the HNLs with the active neutrinos. Instead, cross sections are proportional to Λ^{-4}, where Λ is the new physics scale. At the same time, the pair-N_R operators cannot cause (the lightest) HNL to decay. Thus, HNLs decay only via their mixing with the active neutrinos, $V_{\alpha N}$. This scenario leads to one very important change in phenomenology with respect to previous works, that considered HNLs at the LHC which were produced via charged (and neutral) current processes induced by the mixing of the HNLs only. Namely, the total signal event number for the different experiments in our set-up scales at the smallest $|V_{\alpha N}|^2$ that one can probe only as $|V_{\alpha N}|^2$, instead of the usual $|V_{\alpha N}|^4$.

We have estimated the sensitivity ranges of various LHC experiments in this set-up: ATLAS and a series of proposed far detectors: namely, AL3X, ANUBIS, CODEX-b, FASER, MoEDAL-MAPP and MATHUSLA. Our main result is that for Λ lower than roughly $\Lambda \sim 10$–15 TeV, depending on the operator, much larger HNL masses and much smaller mixing angles, $|V_{\alpha N}|^2$, can be probed than in the minimal case, where both production cross section and decay lengths are determined by the mixing angle (and the HNL mass) only.

Further, we have considered the LNC four-fermion single-N_R operators associated with a charged lepton and two quarks. They induce both HNL production and decay simultaneously. For HNLs with masses of $\mathcal{O}(10)$ GeV, such operators with $\Lambda \gtrsim 1$ TeV can easily make the HNLs become long-lived, leading to displaced vertices at the LHC. We have therefore proposed a displaced-vertex search strategy based on a prompt-lepton trigger and selection of high-quality displaced tracks. We have estimated the sensitivity reach of ATLAS in the HL-LHC era, with an integrated luminosity of 3 ab^{-1}, to four single-N_R operators: $\mathcal{O}_{duNe}$, $\mathcal{O}_{LNQd}$, $\mathcal{O}_{LdQN}$ and $\mathcal{O}_{QuNL}$.

We have considered the operators with the first-generation quarks and a charged lepton. For simplicity, we did not take into account the effect of active-heavy neutrino mixing, which is supposed to be negligible if the type I seesaw relation is assumed. The proposed displaced-vertex search for HNLs at ATLAS can probe new physics scales up to about 20 (15) TeV, and in some cases above, for HNL mass between about 5 GeV and 50 GeV, for the single-N_R operators with an electron (tau lepton).

Acknowledgements I am grateful to the organisers of the 8th Workshop on Theory, Phenomenology and Experiments in Flavour Physics (FPCapri2022) held on Capri, Italy, for the opportunity to present this work in such a beautiful place. I would like to thank R. Beltrán, G. Cottin, J. C. Helo, M. Hirsch and Z. S. Wang for the enjoyable collaboration on the work summarised in this contribution. I would also like to thank M. Chala and A. Santamaria for many useful discussions on EFTs. This work was supported by the "Generalitat Valenciana" under the grant PROMETEO/2019/087, as well as by the FEDER/MCIyU-AEI grant FPA2017-84543-P and the MICINN-AEI (10.13039/501100011033) grant PID2020-113334GB-I00. Finally, I would like to express special thanks to the Mainz Institute for Theoretical Physics (MITP) of the Cluster of Excellence PRISMA+ (Project ID 39083149), for its partial support during the Scientific Programme "Neutrinos, Flavour and Beyond", held on Capri, Italy, from 6–17 June 2022.

References

1. P. Minkowski, Phys. Lett. B **67**, 421 (1977). https://doi.org/10.1016/0370-2693(77)90435-X
2. T. Yanagida, Conf. Proc. C **7902131**, 95 (1979)
3. M. Gell-Mann, P. Ramond, R. Slansky, Conf. Proc. C **790927**, 315 (1979)
4. S.L. Glashow, N.A.T.O. Sci, Ser. B **61**, 687 (1980). https://doi.org/10.1007/978-1-4684-7197-7_15
5. R.N. Mohapatra, G. Senjanovic, Phys. Rev. Lett. **44**, 912 (1980). https://doi.org/10.1103/PhysRevLett.44.912
6. R. Mohapatra, J. Valle, Phys. Rev. D **34**, 1642 (1986). https://doi.org/10.1103/PhysRevD.34.1642

7. J. Bernabeu, A. Santamaria, J. Vidal, A. Mendez, J.W.F. Valle, Phys. Lett. B **187**, 303 (1987). https://doi.org/10.1016/0370-2693(87)91100-2
8. E.K. Akhmedov, M. Lindner, E. Schnapka, J. Valle, Phys. Lett. B **368**, 270 (1996). https://doi.org/10.1016/0370-2693(95)01504-3
9. E.K. Akhmedov, M. Lindner, E. Schnapka, J. Valle, Phys. Rev. D **53**, 2752 (1996). https://doi.org/10.1103/PhysRevD.53.2752
10. I. Brivio, M. Trott, Phys. Rept. **793**, 1 (2019). https://doi.org/10.1016/j.physrep.2018.11.002
11. F. del Aguila, S. Bar-Shalom, A. Soni, J. Wudka, Phys. Lett. B **670**, 399 (2009). https://doi.org/10.1016/j.physletb.2008.11.031
12. A. Aparici, K. Kim, A. Santamaria, J. Wudka, Phys. Rev. D **80**, 013010 (2009). https://doi.org/10.1103/PhysRevD.80.013010
13. Y. Liao, X.D. Ma, Phys. Rev. D **96**(1), 015012 (2017). https://doi.org/10.1103/PhysRevD.96.015012
14. J. Alimena et al., J. Phys. G **47**(9), 090501 (2020). https://doi.org/10.1088/1361-6471/ab4574
15. V.V. Gligorov, S. Knapen, B. Nachman, M. Papucci, D.J. Robinson, Phys. Rev. D **99**(1), 015023 (2019). https://doi.org/10.1103/PhysRevD.99.015023
16. M. Bauer, O. Brandt, L. Lee, C. Ohm (2019). arXiv:1909.13022
17. V.V. Gligorov, S. Knapen, M. Papucci, D.J. Robinson, Phys. Rev. D **97**(1), 015023 (2018). https://doi.org/10.1103/PhysRevD.97.015023
18. J.L. Feng, I. Galon, F. Kling, S. Trojanowski, Phys. Rev. D **97**(3), 035001 (2018). https://doi.org/10.1103/PhysRevD.97.035001
19. A. Ariga et al., Phys. Rev. D **99**(9), 095011 (2019). https://doi.org/10.1103/PhysRevD.99.095011
20. J.L. Pinfold, Universe **5**(2), 47 (2019). https://doi.org/10.3390/universe5020047
21. J.L. Pinfold, Phil. Trans. Roy. Soc. Lond. A **377**(2161), 20190382 (2019). https://doi.org/10.1098/rsta.2019.0382
22. J.P. Chou, D. Curtin, H.J. Lubatti, Phys. Lett. B **767**, 29 (2017). https://doi.org/10.1016/j.physletb.2017.01.043
23. D. Curtin et al., Rept. Prog. Phys. **82**(11), 116201 (2019). https://doi.org/10.1088/1361-6633/ab28d6
24. C. Alpigiani, et al. (2020). arXiv:2009.01693
25. F. Kling, S. Trojanowski, Phys. Rev. D **97**(9), 095016 (2018). https://doi.org/10.1103/PhysRevD.97.095016
26. J.C. Helo, M. Hirsch, Z.S. Wang, JHEP **07**, 056 (2018). https://doi.org/10.1007/JHEP07(2018)056
27. D. Dercks, H.K. Dreiner, M. Hirsch, Z.S. Wang, Phys. Rev. D **99**(5), 055020 (2019). https://doi.org/10.1103/PhysRevD.99.055020
28. M. Hirsch, Z.S. Wang, Phys. Rev. D **101**(5), 055034 (2020). https://doi.org/10.1103/PhysRevD.101.055034
29. J. De Vries, H.K. Dreiner, J.Y. Günther, Z.S. Wang, G. Zhou, JHEP **03**, 148 (2021). https://doi.org/10.1007/JHEP03(2021)148
30. G. Cottin, J.C. Helo, M. Hirsch, A. Titov, Z.S. Wang, JHEP **09**, 039 (2021). https://doi.org/10.1007/JHEP09(2021)039
31. R. Beltrán, G. Cottin, J.C. Helo, M. Hirsch, A. Titov, Z.S. Wang, JHEP **01**, 044 (2022). https://doi.org/10.1007/JHEP01(2022)044
32. S. Weinberg, Phys. Rev. Lett. **43**, 1566 (1979). https://doi.org/10.1103/PhysRevLett.43.1566
33. B. Grzadkowski, M. Iskrzynski, M. Misiak, J. Rosiek, JHEP **10**, 085 (2010). https://doi.org/10.1007/JHEP10(2010)085
34. A. Atre, T. Han, S. Pascoli, B. Zhang, JHEP **05**, 030 (2009). https://doi.org/10.1088/1126-6708/2009/05/030
35. K. Bondarenko, A. Boyarsky, D. Gorbunov, O. Ruchayskiy, JHEP **11**, 032 (2018). https://doi.org/10.1007/JHEP11(2018)032
36. G. Cottin, J.C. Helo, M. Hirsch, Phys. Rev. D **98**(3), 035012 (2018). https://doi.org/10.1103/PhysRevD.98.035012

Chapter 9
Neutrinos and Dark Sectors

Jaime Hoefken Zink and Silvia Pascoli

Abstract Neutrino physics is key to access physics beyond the Standard Model. Not only the discovery of neutrino oscillations implies that the Standard Model of Particles is incomplete but neutrinos are the most elusive particles and could couple to unexplored dark sectors. In this case, new signatures can arise and can be tested in current and future experiments. They could even solve some of the current anomalies in the leptonic sector and offer some alternatives to some of the most challenging problems in particle physics and cosmology, such as dark matter and the baryon asymmetry.

9.1 Introduction

The discovery of neutrino oscillations, and consequently of neutrino masses, has opened a new window on the physics beyond the Standard Model (BSM) with puzzling questions: how do neutrinos acquire mass? Are they Dirac or Majorana fermions? How do they relate to other BSM sectors?

In the following, we will briefly review the current knowledge of neutrino properties, see Sect. 9.2, and of the main explanations for neutrino masses, see Sect. 9.3, from high energy scales as required in GUT theories to scales as low as the eV. Then we will focus on this second possibility and discuss how dark sectors, i.e. extensions of the Standard Model (SM) below the electroweak scale, could help to address some of the current anomalies in the leptonic sector, making reference to a specific dark model, see Sect. 9.4.

J. Hoefken Zink · S. Pascoli (✉)
Dipartimento di Fisica e Astronomia, Università di Bologna, via Irnerio 46, 40126 Bologna, Italy
e-mail: silvia.pascoli@unibo.it

INFN, Sezione di Bologna, viale Berti Pichat 6/2, 40127 Bologna, Italy

G. Ricciardi et al. (eds.), *8th Workshop on Theory, Phenomenology and Experiments in Flavour Physics*, Springer Proceedings in Physics 292,
https://doi.org/10.1007/978-3-031-30459-0_9

9.2 Neutrino Physics and Current Status

9.2.1 Neutrinos in the SM

In the SM, neutrinos are doublets, together with their respective charged leptons, under the $SU(2)_L$ gauge symmetry and come in three families. A fourth active one is not allowed by the invisible width of the Z boson because it would contribute to its invisible decay ($Z \to \nu_\alpha \nu_\alpha$). The number of active neutrino species N_ν [1] is constrained to be:

$$N_\nu = \frac{\Gamma_{\text{inv}}}{\Gamma_{\nu\bar{\nu}}} = 2.984 \pm 0.008 \tag{9.1}$$

Additional neutrinos need not to partake in SM interactions, reason why they are called "sterile", but they could still mix with the active ones.

9.2.2 Neutrino Mixing

Neutrino oscillations successfully explained the solar and atmospheric neutrino "anomalies" and proved that neutrinos have mass. The idea is that there are massive neutrino states, that is, states with definite mass, which differ from the flavour states, that is, the states produced and detected through the weak interactions with the corresponding charged lepton. The mixing is quantified in terms of the so-called Pontecorvo-Maki-Nakagawa-Sakata matrix, U, [2–4] parameterised in terms of 3 angles and one CP-violating phase (for Dirac neutrinos) or three phases (for Majorana neutrinos). The relation between flavour (ν_α) and mass (ν_i) fields is given by:

$$\nu_{\alpha L} = \sum_{i=1}^{3} U_{\alpha i} \nu_{iL} \tag{9.2}$$

This mixing matrix enters the charged-current (CC) Lagrangian of the SM as:

$$\mathcal{L}_{\text{SM}} = -\frac{g}{\sqrt{2}} \sum_{\alpha i} \bar{\nu}_i U^*_{\alpha i} \gamma^\mu P_L l_\alpha W_\mu + \text{h.c.} \,. \tag{9.3}$$

The standard parameterisation of the PMNS matrix is [5, 6]:

$$U_{\alpha i} = \begin{pmatrix} c_{12}c_{13} & s_{12}c_{13} & s_{13}e^{-i\delta} \\ -s_{12}c_{23} - c_{12}s_{23}s_{13}e^{i\delta} & c_{12}c_{23} - s_{12}s_{23}s_{13}e^{i\delta} & s_{23}c_{13} \\ s_{12}s_{23} - c_{12}c_{23}s_{13}e^{i\delta} & -c_{12}s_{23} - s_{12}c_{23}s_{13}e^{i\delta} & c_{23}c_{13} \end{pmatrix} \cdot \mathcal{D} \,, \tag{9.4}$$

where $c_{ij} \equiv \cos\theta_{ij}$, $s_{ij} \equiv \sin\theta_{ij}$ and $\theta_{ij} \in [0, 90°]$. $\delta \in [0, 360°]$ is the Dirac phase and $\mathcal{D} \equiv \text{diag}(1, e^{i\frac{\alpha_{21}}{2}}, e^{i\frac{\alpha_{31}}{2}})$ contains the two Majorana phases. The Dirac phase is responsible for CP violation in oscillations, which can be quantified in terms of the Jarlskog invariant [7]:

$$J \equiv \mathcal{I}[U_{\mu 3}U_{e2}U^*_{\mu 2}U^*_{e3}] = \frac{1}{8}\sin 2\theta_{12}\sin 2\theta_{23}\sin 2\theta_{13}\cos\theta_{13}\sin\delta\ . \qquad (9.5)$$

9.2.3 Neutrino Oscillations

Neutrino oscillations occur when one flavour state is produced and another one is detected after the state has propagated. In addition to the mismatch between flavour and mass states already discussed, neutrino oscillations require a difference in mass of the mass states, so that they propagate with different phases. To quantify this, let's consider a flavour state, $|\nu_\alpha\rangle$, produced at time $t = 0$. It relates to the mass states, $|\nu_i\rangle$, as:

$$|\nu, t = 0\rangle = |\nu_\alpha\rangle = \sum_i U^*_{\alpha i}|\nu_i\rangle\ . \qquad (9.6)$$

Solving the Shrödinger equation with the Hamiltonian in vacuum, $\hat{H}$, the evolved state can be expressed as

$$|\nu, t\rangle = \exp(-i\hat{H}t)|\nu_\alpha\rangle = \sum_i U^*_{\alpha i}\exp(-iE_i t)|\nu_i\rangle\ , \qquad (9.7)$$

where the eigenvalues are $E_i = \sqrt{p^2 + m_i^2}$, for given momentum p.

Since neutrinos are highly relativistic and we can always subtract a common global phase to the evolution of quantum states, the probability of having an oscillation from a flavour α to a flavour β is found to be

$$P(\nu_\alpha \to \nu_\beta, t) = |\langle\nu_\beta|\nu, t\rangle|^2 = \left|\sum_i U_{\beta i}U^*_{\alpha i}\exp\left(-i\frac{\Delta m^2_{i1}L}{2E}\right)\right|^2 \qquad (9.8)$$

where $E_i - E_j \simeq \frac{m_i^2 - m_j^2}{2p} \simeq \frac{\Delta m^2_{ij}}{2E}$ and $t \simeq L$, being L the baseline.

Using the properties of unitary matrices, we can split the formula into a real and an imaginary part of the leptonic mixing terms:

$$\begin{aligned}P_{\nu_\alpha\to\nu_\beta} &= \delta_{\alpha\beta} - 4\sum_{i>j}\mathcal{R}[U^*_{\alpha i}U_{\alpha j}U^*_{\beta j}U_{\beta i}]\sin^2\left(\frac{\Delta m^2_{ij}L}{4E}\right)\\ &+2\sum_{i>j}\mathcal{I}[U^*_{\alpha i}U_{\alpha j}U^*_{\beta j}U_{\beta i}]\sin\left(\frac{\Delta m^2_{ij}L}{2E}\right).\end{aligned} \tag{9.9}$$

We notice that neutrino oscillation respect leptonic number, though not for each flavour separately. We also see that the Majorana phases do not affect the oscillation as expected as they do not break lepton number. The derivation of the probability in Eq. (9.9) is called the plane-wave derivation, which is simple though it does not account for the momentum uncertainty necessary for coherence and the spatial size of the neutrino wave function resulting from the fact that production and detection are localised processes. A deeper derivation of the process could be done using wave packets [8].

For the case of three flavours, the probabilities depend on six independent parameters inherent to the physics of the neutrinos: θ_{12}, θ_{13}, θ_{23}, δ, Δm^2_{21}, Δm^2_{31} and two which depend on the experiment: the energies, E, and the baseline, L. Δm^2_{32} can be directly obtained from the other square mass differences.

In the simplified case of two flavours, there is one angle for the mixing matrix, θ, and one mass square difference, Δm^2 and the probability reduces to

$$\begin{aligned}P_{\nu_\alpha\to\nu_\beta} &= \delta_{\alpha\beta} + (-1)^{\delta_{\alpha\beta}}\sin^2 2\theta\sin^2\left(\frac{\Delta m^2 L}{4E}\right)\\ &= \delta_{\alpha\beta} + (-1)^{\delta_{\alpha\beta}}\sin^2 2\theta\sin^2\left(1.27\frac{\Delta m^2}{\text{eV}^2}\frac{L}{\text{km}}\frac{\text{GeV}}{E}\right).\end{aligned} \tag{9.10}$$

Fitting all available data from neutrino oscillation experiments, the best fits for the oscillation parameters are the following [9] (Table 9.1).

Table 9.1 Current best fits for neutrino oscillation physics. The uncertainties are given at 1σ. The parameter l in Δm^2_{3l} is 1 for normal ordering and 2 for inverted ordering

Best fits for oscillation parameters

Parameter	Normal ordering (best fit)	Inverted ordering ($\Delta\chi^2 = 2.6$)
$\theta_{12}/°$	$33.45^{+0.012}_{-0.012}$	$33.45^{+0.78}_{-0.75}$
$\theta_{23}/°$	$42.1^{+1.1}_{-0.9}$	$49.0^{+0.9}_{-1.3}$
$\theta_{13}/°$	$8.62^{+0.12}_{-0.12}$	$8.61^{+0.14}_{-0.12}$
$\delta/°$	230^{+36}_{-25}	278^{+22}_{-30}
$\frac{\Delta m^2_{21}}{10^{-5}\ \text{eV}^2}$	$7.42^{+0.21}_{-0.20}$	$7.42^{+0.21}_{-0.20}$
$\frac{\Delta m^2_{3l}}{10^{-3}\ \text{eV}^2}$	$+2.510^{+0.027}_{-0.027}$	$-2.490^{+0.026}_{-0.028}$

This fitting was done considering the latest neutrino oscillation results from the Neutrino2020 conference, combined with the atmospheric neutrino data analysis performed by the Super-Kamiokande collaboration. With respect to previous fittings, this one is mainly affected by T2K, NOνA and long-baseline experiments. There are also updated analyses from the reactor experiments Double-Chooz and RENO which have been taken into account.

9.2.4 Current Knowledge

As $|\Delta m^2_{31}| \gg \Delta m^2_{21}$, there are two possible ways to arrange the masses. Thanks to matter effects in the Sun, we know that $\Delta m^2_{21} > 0$. On the other hand, Δm^2_{31} could be positive or negative, mass patterns called *normal* or *inverted ordering*, respectively. In term of these two parameters and the minimum mass, $m_{\rm min}$, we can express all the three masses as (i) for normal ordering $m_1 = m_{\rm min}$, $m_2 = \sqrt{m^2_{\rm min} + \Delta m^2_{21}}$, $m_3 = \sqrt{m^2_{\rm min} + \Delta m^2_{31}}$, and (ii) for inverted ordering, $m_3 = m_{\rm min}$, $m_1 = \sqrt{m^2_{\rm min} + \Delta m^2_{32} - \Delta m^2_{21}}$, $m_2 = \sqrt{m^2_{\rm min} + \Delta m^2_{32}}$. We still need to measure $m_{\rm min}$ and discover the ordering in order to know the masses of each neutrino.

The other key unknown is the Dirac phase, δ. Leptonic CP violation is present if $\delta \neq 0, \pi$ and/or the Majorana phases take CP-violating values. CP violation, together with the other Sakharov conditions [10], plays a key role in explaining the baryon asymmetry and can be related to the origin of the flavour structure.

9.2.5 Open Questions

There are crucial open questions regarding neutrino physics which still need to be answered:

1. Are neutrinos Dirac or Majorana fermions?
2. What are the precise values of their masses?
3. Is there any leptonic CP-violation? What is the value of the δ CP violating phase?
4. What are the values of the mixing angles? Are they related to any underlying flavour principle?
5. Is the standard theory of 3-neutrino mixing correct or are there other exotic effects, such as sterile neutrinos, non-standard interactions, Lorentz-violation?

A wide experimental program is addressing these questions and will answer many of them in the next decade. Neutrinoless double beta decay is the most sensitive process to whether neutrinos are of Majorana nature or Dirac and would also give information on their absolute mass spectrum. Accelerator neutrino experiments, as well as atmospheric ones, such as DUNE, HK, KM3Net, IceCube could determine the mass ordering by taking into account matter effects in neutrino oscillations, with DUNE reaching 5 σ discovery in about a decade from now. This information can also be provided by the medium baseline reactor neutrino experiment JUNO. Cosmology sets the most stringent bounds on the sum of neutrino masses, albeit within the standard cosmological framework, and might measure them with future observations. There are also direct neutrino mass searches which exploits the effect of those on the energy spectrum of beta decay generally from tritium, such as KATRIN. Long baseline neutrino oscillation experiments, such as DUNE and T2HK, are also aiming to establish leptonic CP violation by observing the $\nu_\mu \to \nu_e$ transition channel. If the δ phase is close to being maximally violating, they will be able to discover leptonic CP-violation within ten years of running at the nominal exposure. They will also improve the important precise measurement of the angle θ_{23} while JUNO will give the most accurate measurement of θ_{12}. Finally, there are short baseline neutrino oscillation experiments, such as the SBN program at Fermilab, and reactor and radioactive source neutrino experiments searching for sterile neutrinos. Other deviations from the standard 3-neutrino picture are also actively searched for now and in the coming future. Some hints might have been reported but remain controversial and a coherent picture is missing.

9.3 Neutrino Masses BSM

9.3.1 Neutrino Masses Beyond the SM

In the SM, neutrinos do not acquire mass. Therefore, the way we need to extend the SM Lagrangian so that they are massive is an key open question.

Dirac Masses If we add right-handed neutrino fields, ν_R, which are SM gauge singlets, we could build the Lagrangian mass terms for neutrinos just as we have for other fermions. The following Yukawa coupling is allowed by the gauge symmetries:

$$\mathcal{L} \supset -y_\nu \bar{L} \cdot \widetilde{H} \nu_R + \text{h.c.}\,, \tag{9.11}$$

where $L \equiv (\nu_L^T, l^T)^T$ is the leptonic left doublet and $\widetilde{H} \equiv i\sigma_2 H^*$. The Dirac mass for light neutrinos is generated when the Higgs acquires a vacuum expectation value different from zero: $\langle \widetilde{H} \rangle = (v_H/\sqrt{2}, 0)^T$. A Dirac mass term arises:

$$\mathcal{L}_D = -\frac{v_H y_\nu}{\sqrt{2}} \bar{\nu}_L \nu_R + \text{h.c.}\,. \tag{9.12}$$

We notice that this explanation constitutes a major departure from the SM as it is necessary to impose lepton number conservation to forbid a Majorana mass term for ν_R which otherwise should be included. This requires to promote lepton number, which is an accidental symmetry of the SM, into a fundamental symmetry of Nature.

Since the masses of the neutrinos are sub-eV, we obtain that $y_\nu \sim 10^{-12}$. We do not have an answer to why this value should be so small compared to the other Yukawa couplings. Moreover, the neutrino mass hierarchy is at most mild and leptonic mixing has a structure which is very different from that of the quark sector. This have suggested other explanations of neutrino masses.

Majorana Masses A Majorana mass term for neutrinos is of the form $\nu_L^T C^\dagger \nu_L$. It is neutral but breaks the SM gauge invariance. However, it would arise from a dimension 5 operator, typically named the Weinberg operator [11]. Using the fact that the term $\bar{L} \cdot \widetilde{H}$ is gauge invariant, it given by

$$\mathcal{L}_M = -\frac{\lambda}{\Lambda} L^T \cdot \widetilde{H}^* C^\dagger \widetilde{H}^\dagger \cdot L + \text{h.c.} \xrightarrow{\langle \widetilde{H} \rangle \neq 0} -\frac{\lambda v_H^2}{2\Lambda} \nu_L^T C^\dagger \nu_L + \text{h.c.}\,. \tag{9.13}$$

This operator breaks lepton number by to units and produces the Majorana mass term after the Higgs boson gets a non-vanishing vacuum expectation value.

We notice that it requires a mass scale Λ in the denominator. This suggests that there is a new theory at a high scale Λ which is integrated out at low energies, such as three main types of see-saw mechanisms, as well as extended versions, such as the inverse, linear and extended see-saw. Here, we will focus on the most studied case, the see-saw type I mechanism [12–15].

See-Saw Type I Model This is the simplest model that can account for neutrino masses and their smallness. As lepton number is broken, it predicts Majorana neutrino masses. We will consider N sterile neutrinos, $N_{j,R}$, with $1 \le j \le N$. We can write the following general Lagrangian that respects the SM gauge group:

$$\mathcal{L}_{ss} = \mathcal{L}_{SM} - \sum_{j,\alpha} y_{\alpha j} \bar{L}_\alpha \cdot \widetilde{H} N_{j,R} + \sum_{j,k} \frac{1}{2} N_{j,R}^T C^\dagger M_{N,jk} N_{k,R} + \text{h.c.}\,, \tag{9.14}$$

where y is a $3 \times j$ matrix and M_N is a $j \times j$ symmetric Majorana mass matrix, which can be taken to be real and diagonal, with heavy masses M_j. After the Higgs boson gets a vev, the Yukawa term will induce Dirac mass terms for neutrinos: $m_D \equiv y v_H/\sqrt{2}$. So, the mass terms read:

$$\mathcal{L}_{ss,mass} = \frac{1}{2} \left((\nu_L^c)^T \ N_R^T \right) C^\dagger \begin{pmatrix} 0 & m_D \\ m_D^T & M_N \end{pmatrix} \begin{pmatrix} \nu_L^c \\ N_R \end{pmatrix} + \text{h.c.}\,. \tag{9.15}$$

We consider the limit in which $m_D \ll M_N$, so that after diagonalizing the mass matrix the heavy neutrinos are mainly sterile with a mass close to M_j, while the light neutrinos are mainly active with a small mass due the suppression by a large M_N:

$$m_\nu \simeq -m_D \frac{1}{M_N} m_D^T\,. \tag{9.16}$$

The scale of the right handed neutrinos is not known and can span many orders of magnitude going from scales as high as GUT theories to the eV, e.g. for low energy see-saw models. In the first case, the see-saw type mechanism can be easily implemented in GUT models [16–18], e.g. $SO(10)$ which already include right-handed neutrinos, and can lead to the baryon asymmetry via leptogenesis [19–24], and could even be connected to the production of gravitational waves [25–28]. These models are very difficult to test as the new particles are too heavy to be accessed directly at experiments. In recent years more and more attention has been devoted to models at lower scales that, in principle, could be tested directly. We will focus on this option in the next section.

9.4 Dark Sectors

The most important evidence that the Standard Model (SM) of particle physics is incomplete comprises neutrino masses and mixing, and the presence of dark matter (DM) and of a baryon asymmetry in the Universe. Both call for extensions of the SM. While for many years the focus has been on high energies, recently great attention has been given to dark sectors at scales below the electroweak one. They do not partake in SM interactions, or do so with extremely weak couplings while they can still display strong "dark" interactions. The dark sector can interact with the SM via the so-called "portals", renormalizable interactions that are classified in (i) vector portal when in a $U(1)_D$ gauge symmetry extension of the SM, the field X^μ is kinetically mixes with the hypercharge field or the photon. The resulting boson is called a "dark photon", (ii) scalar portal when a dark Higgs singlet mixes with the SM one, (iii) the neutrino portal when new right-handed neutrinos couple on one side with SM neutrinos and on the other with the dark sector. In this case the right handed neutrinos which mix with the SM ones go under the name of heavy neutral leptons (HNL). For a review of dark sectors theory and phenomenology, we refer the reader to [29–32].

9.4.1 Dark Model

Dark sector models could be minimal, if a very restricted number of new particles and interactions are considered, e.g. in the case of a single portal, or "rich" if multiple particles and/or generations are accounted for. The latter case is strongly motivated by analogy with the structure of the SM and by the self-consistency of the model, e.g. dark photons and HNLs require a mechanism for their masses.

Here, we review the three-portal model [30], as a typical example of "rich" dark sector with unique phenomenological signatures. The model has a $U(1)$ extension of the SM, with a dark scalar which breaks the symmetry giving mass to the dark photon and to the HNLs, and a set of N_R and dark fermions, charged under the gauge symmetry. The dark photon kinetically mixes with the hypercharge boson.

The Lagrangian of the model is the following:

$$\begin{aligned}\mathcal{L} \supset &-\frac{1}{4}X^{\mu\nu}X_{\mu\nu} - \frac{\sin\chi}{2}X_{\mu\nu}B^{\mu\nu} + (\mathcal{D}^{x}_{\mu}\Phi)^{\dagger}(\mathcal{D}^{\mu}_{x}\Phi)\\ &- V(\Phi, H) + \overline{\nu_N} i\partial\nu_N + \overline{\nu_D} i\mathcal{D}^{x}\nu_D\\ &-\Bigg[(\overline{L}\widetilde{H})Y\nu_N^c + \frac{1}{2}\nu_N M_N \nu_N^c + \overline{\nu_N}(Y_L \nu_{D_L}^c \Phi + Y_R \nu_{D_R}\Phi^*)\\ &+ \overline{\nu}_D M_X \nu_D + \text{h.c.}\Bigg]\,. \end{aligned} \tag{9.17}$$

where $D^{x}_{\mu}\Phi \equiv \partial_\mu - i g_X X_\mu$.

After the dark and the electro-weak symmetry breaking, it is possible to write the neutrino mass matrix in terms of the flavour states $\nu_f = (\nu_\alpha^c \;\; \nu_N^c \;\; \nu_{D_L}^c \;\; \nu_{D_R})^T$, where ν_α, with $\alpha = \{e, \mu, \tau\}$ are the SM neutrinos flavour states. The dark Higgs takes a vev at the GeV scale, which is also the scale for the new broken vector field that arises, the dark photon, Z': $v_\phi, m_{Z'} \sim$ GeV.

After defining $\epsilon = \chi \times \cos\theta_W$, the main interactions of the dark photon are

$$-\mathcal{L}_{Z'}^{(\text{int})} \supset \epsilon e Z'_\mu J^\mu_{(\text{EM})} + g_D U^*_{Di} U_{Dj} Z'_\mu J^\mu_{(\nu)}\,, \tag{9.18}$$

where $J^\mu_{(\text{EM})}$ and $J^\mu_{(\nu)}$ stand for the electromagnetic and the neutrino currents respectively. Therefore, fermion currents that interact through the photon or neutrino currents can both interact through the dark photon, as we can see from the coupled fields in Eq. (9.18). That means that neutrinos can interact with ordinary matter through the dark photon.

The phenomenology of such model depends critically on the hierarchy of the masses of the new particles, on the couplings and values of the masses and could be very varied. As a specific example, we present the parameters discussed in [30] as this offers an explanation for various anomalies, such as the MiniBooNE low energy excess (LEE), the value of $g-2$ of the muon. We consider a GeV scale for the new physics. Specifically, we fix the dark photon mass around 1.25GeV and assume that at least 3 of the massive HNLs, N_4, N_5, N_6, are lighter than the dark photon.

In this case, the dark photon mainly decays into HNL pairs and, for the specific benchmark point considered, the branching ratios in $Z' \to N_i N_j$ are given in the Table 9.2.

Table 9.2 Branching ratios of different decay modes of the dark photon. Data taken from [32]

$\mathcal{B}(Z' \longrightarrow N_i N_j)/\%$					
44	45	46	55	56	66
0.15	11	0.48	1.6	86	0.59

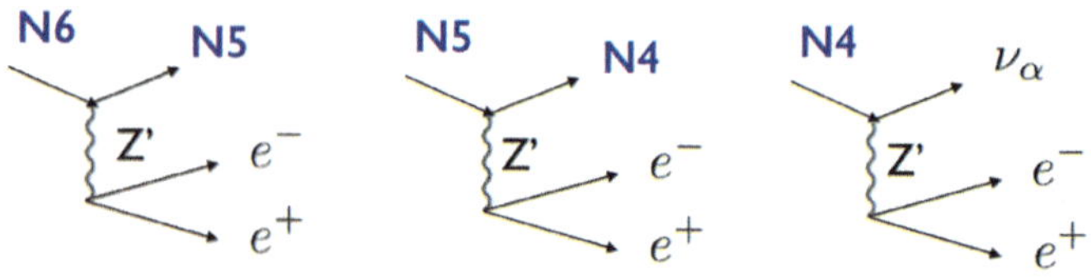

Fig. 9.1 Diagrams of the decay of HNLs through the dark photon

We are regarding HNLs of masses of the order of 100*s*MeV. They can mainly decay through the dark photon into lighter neutrino states and an electron-positron pair with decay rates much faster than in the SM (plus HNL mixing) (Fig. 9.1).

This would be a major departure from "traditional" BSM thinking through the opening of three portals that are susceptible of being detected in current experiments. One can expect, for example, displaced vertices and decay chains, for which MicroBooNE, DUNE-ND and T2K would have excellent sensitivity. Also NA62 and SHADOWS could search for signals in kaon decays, while Belle II and BES III can look for signatures in dark photon searches through both invisible and semi-visible decays.

9.4.2 The MiniBooNE Low Energy Excess

The MiniBooNE experiment sits on the Fermilab Booster Neutrino Beamline (BNB) and employs a 12.2 m diameter Cherenkov detector filled with 818 tons of pure mineral oil, located 541 m away from a beryllium target. It ran in both neutrino mode, with a forward-horn-current (FHC), and anti-neutrino mode, with a reverse-horn-current (RHC). Over a span of 17 years, MiniBooNE observed a large low-energy excess (LEE) of electron-like events. The first excess was reported in neutrino mode, between the years 2007–2009. For a total of 6.46×10^{20} Protons-on-target (POT), an excess of 128.8 ± 43.4 electron-like events was observed over the background with a significance of 3.0σ [33]. This excess was predominantly present in the $200\text{MeV} < E_\nu^{QE} < 475\text{MeV}$ energy region. It was subsequently observed that an excess of comparable significance (2.8σ) was also present in the anti-neutrino mode, with 78.4 ± 28.5 excess events observed over the background between the energies $200\text{MeV} < E_\nu^{QE} < 1250\text{MeV}$ and for 11.27×10^{20} POT [34, 35]. Following these tantalizing hints of new physics, the collaboration increased the data in neutrino mode, nearly doubling the POT, while improving the background analysis and reducing the systematic uncertainties. This led to a substantial increase in the significance of the observations, with the $\sim 3\sigma$ excess rising to 4.7σ [36], and most recently to 4.8σ [37]. The combined neutrino and anti-neutrino mode excess currently stands at $638.0 \pm 52.1(\text{stat.}) \pm 122.2(\text{syst.})$ electron-like events (Fig. 9.2).

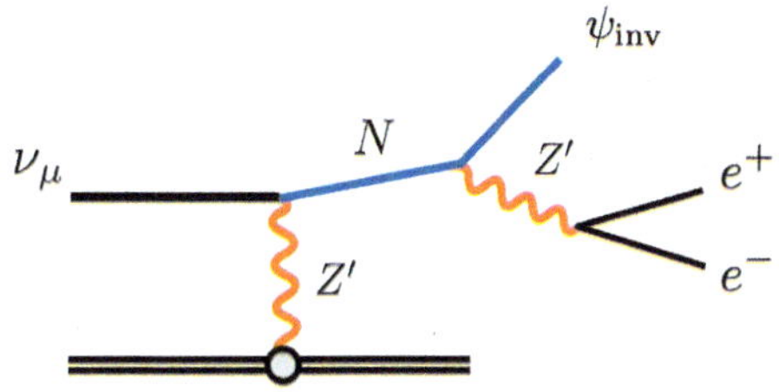

Fig. 9.2 Neutrino upscattering to HNLs and subsequent decay of the latter to electron-positron pairs plus invisible states

The striking 4.8σ excess of low-energy electron-like events (LEE) observed at MiniBooNE [33, 35, 38] challenges our understanding of the neutrino sector at GeV energies. This excess has been scrutinized throughout the literature, with some authors pointing to combinations of effects that could reduce its significance to the order of $3\,\sigma$ [1, 39]. At this time, it is fair to say that the origin of the LEE remains utterly unknown.

A new generation of experiments with improved particle-identification (PID) capabilities has started to test the origin of the LEE. The first tests came from the MicroBooNE experiment, which used a 170-ton Liquid Argon Time Projection Chamber (LArTPC) in the same beam, the Booster Neutrino Beamline (BNB) as MiniBooNE. Two additional detectors will add to the Short-Baseline Neutrino (SBN) program, the Icarus detector, a 760 ton detector located 600 m from the BNB target, and the Short-Baseline Near Detector (SBND), with a 112-ton active volume at a shorter distance, 110 m away.

Possible Explanations of MB LEE The most studied explanation of the MiniBooNE LEE anomaly is within the in-vacuum $(3+1)$-neutrino oscillations, induced by an eV-scale sterile neutrino. However, this paradigm faces strong tensions with cosmology [40–42] and ν_μ-disappearance data [43–49]. Recent data from MicroBooNE shows no excess of ν_e events [50–53], but are thus far unable to fully exclude the $3+1$-oscillation picture [54–56] (see also [57]). It is, however, fair to conclude that explanations based on $3+1$-neutrino oscillations, despite their simplicity, are not the forefront contenders to explain the MiniBooNE excess nor the global short-baseline data. There are many other possibilities on how to explain the anomaly. Since MiniBooNE uses a Cherenkov detector, the LEE can be mimicked by multiple kinds of electromagnetic final states, such as electrons, photons, or combinations thereof.

On the one hand, single electrons have been disfavored by MicroBooNE analyses. These final states are not capable of explaining such an excess for most of the topologies which MicroBooNE can distinguish [50]. On the other hand, single photons due to SM processes have also been disfavored by the last analyses [58]. A considerable unseen excess of photons would have been detected in MicroBooNE if they were responsible of explaining the LEE at MiniBooNE. Therefore, more exotic explanations could be hidden behind this anomaly.

Highly collimated electron-positron/photon pairs or very asymmetric in energy could be misreconstructed as single electron-like events in MiniBooNE. So many scenarios beyond the SM might explain this excess. Among them, dark sectors play

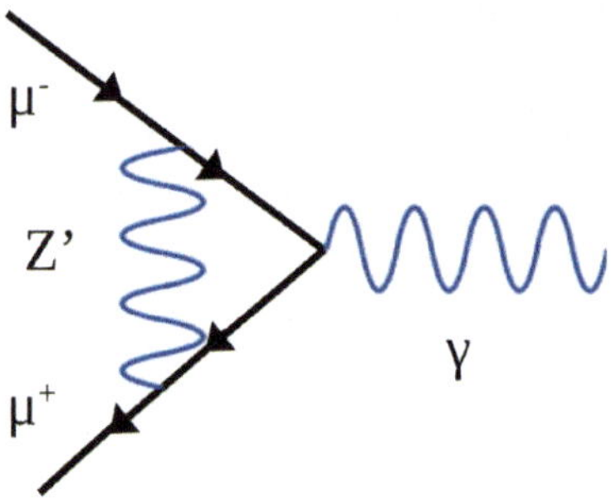

Fig. 9.3 Contribution of the dark photon to the anomalous magnetic moment of the muon

an important role [32, 59] and they could even be tested using sophisticated simulators [60]. Dark models include extended Higgs sectors with decay of scalars [61, 62], decays of leptophilic axion-like particles [63] or decays of new dark gauge bosons [29, 30, 32, 59], whose model has been explained in Sect. 9.4.1.

Considering the decay of dark photons, an incoming muon neutrino could interact with the material of the detector to upscatter to a HNL, which subsequently would decay through a dark photon in the detector, emitting an electron-positron pair [59]. These two showers, when collimated or asymmetric, could be responsible of the LEE. Originally this model inspired on the explanation, now excluded, given in [64], through the decay of a HNL produced from a SM neutral current interaction, subsequently decaying into a photon and a neutrino. A similar analysis was developed in [29], but just considering a light dark photon which give a too-forward spectrum and is highly constrained by other searches at neutrino experiments [31, 65, 66]. A heavier dark photon, $\mathcal{O}(1\text{GeV})$, was proposed in [59]. It was subsequently pointed out that multiple HNLs can lead to faster decays and a better fit in reproducing the LEE [32, 59]. This last reference presents a fit for the energy of the showers in MiniBooNE, considering a dark photon of 1.25GeV and the decay channels: $N_5 \to N_4 e^+ e^-$ and $N_6 \to N_4 e^+ e^-$.

The bounds for the mixing parameters of these models are really altered if additional interactions are allowed, so that HNLs can decay invisibly or semivisibly [31]. Potentially, strong bounds come from ND280 in T2K.

$(\mathbf{g}-\mathbf{2})_\mu$ Anomaly and Dark Sectors There is a longstanding discrepancy between the measured value of $(g-2)_\mu$ and the theoretical prediction at 4.2 σ. Given that $a_\mu \equiv (g-2)_\mu/2$, then $\Delta a_\mu \equiv a_\mu^{\text{exp}} - a_\mu^{\text{th}} = (274 \pm 73) \times 10^{-11}$ [67] (Fig. 9.3).

One way to explain this anomaly is through contributions from a dark photon: whether emerging from super-symmetric extensions [68] or just as a sector with weak admixture to photons [69], or by kinetic mixing and a light vector broken symmetry. The dark boson contribution, which is presented in Fig. 9.3, could explain this discrepancy. Taken into account current constraints, this explanation is allowed if dark photons decay semivisibly, i.e. they decay fast into visible particle and some missing energy [32, 70]. If the missing energy is not too much, collider bounds, in particular from BaBar, can be relaxed and a viable region opens up at the $\mathcal{O}(1\text{GeV})$ range [32], in the same range needed to explain the MiniBooNE LEE.

9.5 Conclusions

Neutrino physics offers the only particle physics evidence of new physics BSM. The hunt for the origin of neutrino masses and leptonic mixing is one of the most compelling physics questions at present. In the current article, we have reviewed the current status of neutrino physics and the open questions which remain open. We have then discussed how to extend the SM to include neutrino masses, invoking scales which can be as high as GUT ones or as low as few eV. We have put emphasis on physics coming from light sectors, so-called "dark sectors", that can be probed by a very wide experimental programme. In particular, we have highlighted that "rich dark sectors" which comprise multiple particles and interactions in the dark sector, offer unique phenomenology and could explain some of the anomalies that are still unsolved, such as the MiniBooNE low energy excess and the $(g-2)_\mu$.

Acknowledgements The research of J.H.Z. and S.P. has received partial support from the European Union's Horizon 2020 research and innovation programme under the Marie Skłodowska-Curie grant agreement No 860881-HIDDeN. S.P. has also been supported by the MITP Scientific Program "Neutrinos, Flavour and Beyond" on the island of Capri.

References

1. M. Tanabashi et al., Review of particle physics. Phys. Rev. D **98**, 030001 (2018)
2. Bruno Pontecorvo, Mesonium and antimesonium. Sov. J. Exp. Theor. Phys. **6**, 429 (1958)
3. B. Pontecorvo, Inverse β processes and nonconservation of lepton charge. Zhur. Eksptl'. i Teoret. Fiz. **34** (1958)
4. Ziro Maki, Masami Nakagawa, Shoichi Sakata, Remarks on the unified model of elementary particles. Prog. Theor. Phys. **28**(5), 870–880 (1962)
5. S.M. Bilenky, J. Hošek, S.T. Petcov, On the oscillations of neutrinos with Dirac and Majorana masses. Phys. Lett. B **94**(4), 495–498 (1980)
6. J. Schechter, J.W.F. Valle, Neutrino masses in SU (2) U (1) theories. Phys. Rev. D **22**(9), 2227 (1980)
7. Cecilia Jarlskog, Commutator of the quark mass matrices in the standard electroweak model and a measure of maximal CP nonconservation. Phys. Rev. Lett. **55**(10), 1039 (1985)
8. Shmuel Nussinov, Solar neutrinos and neutrino mixing. Phys. Lett. B **63**(2), 201–203 (1976)
9. I. Esteban, M. Conceptión González-García, M. Maltoni, T. Schwetz, A. Zhou, The fate of hints: updated global analysis of three-flavor neutrino oscillations. J. High Energy Phys. **2020**(9), 1–22 (2020)
10. A.D. Sakharov, Violation of CP-invariance, C-asymmetry, and baryon asymmetry of the Universe, in *In The Intermissions... Collected Works on Research into the Essentials of Theoretical Physics in Russian Federal Nuclear Center, Arzamas-16* (World Scientific, 1998), pp. 84–87
11. Steven Weinberg, Baryon-and lepton-nonconserving processes. Phys. Rev. Lett. **43**(21), 1566 (1979)
12. Peter Minkowski, $\mu \rightarrow e\gamma$ at a rate of one out of 109 muon decays? Phys. Lett. B **67**(4), 421–428 (1977)
13. M. Gell-Mann, P. Ramond, R. Slansky, Supergravity, in *vol. 315*. ed. by P. van Nieuwenhuizen, D.Z. Freedman (North-Holland, Amsterdam, 1979), pp.79–18
14. T Yanagida, Proceeding Workshop on Unified theory and the baryon number in the universe. *KEK Report No. 79-18*, 95 (1979)

15. R.N. Mohapatra, G. Senjanović, Neutrino mass and spontaneous parity nonconservation. Phys. Rev. Lett. **44**(14), 912 (1980)
16. Arthur Hebecker, John March-Russell, The flavour hierarchy and see-saw neutrinos from bulk masses in 5d orbifold GUTs. Phys. Lett. B **541**(3–4), 338–345 (2002)
17. L. Calibbi, Y. Mambrini, S. Kumar Vempati, SUSY-GUTs, SUSY-seesaw and the neutralino dark matter. J. High Energy Phys. **2007**(09), 081 (2007)
18. R. Lal Awasthi, M.K Parida, Inverse seesaw mechanism in nonsupersymmetric $SO(10)$, proton lifetime, nonunitarity effects, and a low-mass Z' boson. Phys. Rev. D **86**(9), 093004 (2012)
19. EKh. Akhmedov, M. Frigerio, A. Yu Smirnov, Probing the seesaw mechanism with neutrino data and leptogenesis. J. High Energy Phys. **2003**(09), 021 (2003)
20. C.H. Albright, S.M. Barr, Leptogenesis in the type III seesaw mechanism. Phys. Rev. D **69**(7), 073010 (2004)
21. Thomas Hambye, Goran Senjanović, Consequences of triplet seesaw for leptogenesis. Phys. Lett. B **582**(1–2), 73–81 (2004)
22. M. Drewes, B. Garbrecht, D. Gueter, J. Klarić, Testing the low scale seesaw and leptogenesis. Phys. Lett. B **2017**(8), 1–52 (2017)
23. J.A. Dror, T. Hiramatsu, K. Kohri, H. Murayama, G. White, Testing the seesaw mechanism and leptogenesis with gravitational waves. Phys. Rev. Lett. **124**(4), 041804 (2020)
24. Zhi-zhong Xing, Zhen-hua Zhao, The minimal seesaw and leptogenesis models. Phys. Rev. Lett. **84**(6), 066201 (2021)
25. Wilfried Buchmuller, Valerie Domcke, Hitoshi Murayama, Kai Schmitz, Probing the scale of grand unification with gravitational waves. Phys. Lett. B **809**, 135764 (2020)
26. S.F. King, S. Pascoli, Y.-L. Zhou, J. Turner, Confronting SO (10) GUTs with proton decay and gravitational waves. J. High Energy Phys. **2021**(10), 1–38 (2021)
27. S.F. King, S. Pascoli, J. Turner, Y.-L. Zhou, Gravitational waves and proton decay: complementary windows into grand unified theories. Phys. Rev. Lett. **126**(2), 021802 (2021)
28. D.I. Dunsky, A. Ghoshal, H. Murayama, Y. Sakakihara, G. White, GUTs, hybrid topological defects, and gravitational waves. Phys. Rev. D **106**(7), 075030 (2022)
29. E. Bertuzzo, S. Jana, P.A.N. Machado, R. Zukanovich Funchal, Dark neutrino portal to explain MiniBooNE excess. Phys. Rev. Lett. **121**(24), 241801 (2018)
30. Peter Ballett, Matheus Hostert, Silvia Pascoli, Dark neutrinos and a three portal connection to the standard model. Phys. Rev. D **101**(11), 115025 (2020)
31. Peter Ballett, Tommaso Boschi, Silvia Pascoli, Heavy neutral leptons from low-scale seesaws at the DUNE near detector. J. High Energy Phys. **2020**, 3 (2020)
32. Asli Abdullahi, Matheus Hostert, Silvia Pascoli, A dark seesaw solution to low energy anomalies: MiniBooNE, the muon (g-2), and BaBar. Phys. Lett. B **820**, 136531 (2021)
33. A.A. Aguilar-Arevalo et al., Unexplained excess of electron-like events from a 1-GeV neutrino beam. Phys. Rev. Lett. **102**, 101802 (2009)
34. A.A. Aguilar-Arevalo et al., Event excess in the MiniBooNE search for $\bar{\nu}_\mu \rightarrow \bar{\nu}_e$ oscillations. Phys. Rev. Lett. **105**, 181801 (2010)
35. A.A. Aguilar-Arevalo et al., Improved search for $\bar{\nu}_\mu \rightarrow \bar{\nu}_e$ oscillations in the MiniBooNE experiment. Phys. Rev. Lett. **110**, 161801 (2013)
36. A.A. Aguilar-Arevalo et al., Significant excess of ElectronLike events in the MiniBooNE short-baseline neutrino experiment. Phys. Rev. Lett. **121**(22), 221801 (2018)
37. A.A. Aguilar-Arevalo et al., Updated MiniBooNE neutrino oscillation results with increased data and new background studies. Phys. Rev. D **103**(5), 052002 (2021)
38. A.A. Aguilar-Arevalo et al., A search for electron neutrino appearance at the $\Delta m^2 \sim 1eV^2$ scale. Phys. Rev. Lett. **98**, 231801 (2007)
39. Vedran Brdar, Joachim Kopp, Can standard model and experimental uncertainties resolve the MiniBooNE anomaly? Phys. Rev. D **105**(11), 115024 (2022)
40. M. Tanabashi et al., Review of particle physics. Phys. Rev. D **98**, 030001 (2018)
41. J. Hamann, S. Hannestad, G.G. Raffelt, Y.Y.Y. Wong, Sterile neutrinos with eV masses in cosmology: how disfavoured exactly? JCAP **09**, 034 (2011)

42. Maria Archidiacono, Nicolao Fornengo, Carlo Giunti, Steen Hannestad, Alessandro Melchiorri, Sterile neutrinos: cosmology versus short-baseline experiments. Phys. Rev. D **87**(12), 125034 (2013)
43. Steffen Hagstotz, Pablo F. de Salas, Stefano Gariazzo, Martina Gerbino, Massimiliano Lattanzi, Sunny Vagnozzi, Katherine Freese, Sergio Pastor, Bounds on light sterile neutrino mass and mixing from cosmology and laboratory searches. Phys. Rev. D **104**(12), 123524 (2021)
44. M. Dentler, Á. Hernández-Cabezudo, J. Kopp, P.A.N. Machado, M. Maltoni, I. Martinez-Soler, T. Schwetz, Updated global analysis of neutrino oscillations in the presence of eV-scale sterile neutrinos. JHEP **08**, 010 (2018)
45. A. Diaz, C.A. Argüelles, G.H. Collin, J.M. Conrad, M.H. Shaevitz, Where are we with light sterile neutrinos? Phys. Rept. **884**, 1–59 (2020)
46. P. Adamson et al., Improved constraints on sterile neutrino mixing from disappearance searches in the MINOS, MINOS+, Daya Bay, and Bugey-3 experiments. Phys. Rev. Lett. **125**(7), 071801 (2020)
47. M.G. Aartsen et al., eV-Scale Sterile Neutrino Search Using Eight Years of Atmospheric Muon Neutrino Data from the IceCube Neutrino Observatory. Phys. Rev. Lett. **125**(14), 141801 (2020)
48. M.G. Aartsen et al., Searching for eV-scale sterile neutrinos with eight years of atmospheric neutrinos at the IceCube Neutrino Telescope. Phys. Rev. D **102**(5), 052009 (2020)
49. Basudeb Dasgupta, Joachim Kopp, Sterile Neutrinos, Phys. Rept. **928**, 63 (2021)
50. J.M. Berryman, P. Coloma, P. Huber, T. Schwetz, A. Zhou, Statistical significance of the sterile-neutrino hypothesis in the context of reactor and gallium data. 11 (2021)
51. P. Abratenko et al., Search for an excess of electron neutrino interactions in MicroBooNE using multiple final state topologies. 10 (2021)
52. P. Abratenko et al., Search for an anomalous excess of charged-current quasi-elastic ν_e interactions with the MicroBooNE experiment using Deep-Learning-based reconstruction. 10 (2021)
53. P. Abratenko et al., Search for an anomalous excess of charged-current ν_e interactions without pions in the final state with the MicroBooNE experiment. 10 (2021)
54. P. Abratenko et al., Search for an anomalous excess of inclusive charged-current ν_e interactions in the MicroBooNE experiment using Wire-Cell reconstruction 10 (2021)
55. C.A. Argüelles, I. Esteban, M. Hostert, K.J. Kelly, J. Kopp, P.A.N. Machado, I. Martinez-Soler, Y.F. Perez-Gonzalez, MicroBooNE and the ν_e interpretation of the MiniBooNE low-energy excess. 11 (2021)
56. A.A. Aguilar-Arevalo et al., MiniBooNE and MicroBooNE joint fit to a 3+1 sterile neutrino scenario. 1 (2022)
57. P. Abratenko et al., First constraints on light sterile neutrino oscillations from combined appearance and disappearance searches with the MicroBooNE detector. 10 (2022)
58. P.B. Denton, Sterile neutrino searches with MicroBooNE: electron neutrino disappearance. 11 (2021)
59. P. Abratenko et al., Search for neutrino-induced neutral current Δ radiative decay in MicroBooNE and a first test of the MiniBooNE low energy excess under a single-photon hypothesis. 10 (2021)
60. Peter Ballett, Silvia Pascoli, Mark Ross-Lonergan, U(1)' mediated decays of heavy sterile neutrinos in MiniBooNE. Phys. Rev. D **99**, 071701 (2019)
61. A.M. Abdullahi, J. Hoefken Zink, M. Hostert, D. Massaro, S. Pascoli, DarkNews: a Python-based event generator for heavy neutral lepton production in neutrino-nucleus scattering. 7 (2022)
62. Bhaskar Dutta, Sumit Ghosh, Tianjun Li, Explaining $(g-2)_{\mu,e}$, the KOTO anomaly and the MiniBooNE excess in an extended Higgs model with sterile neutrinos. Phys. Rev. D **102**(5), 055017 (2020)
63. Waleed Abdallah, Raj Gandhi, Samiran Roy, Two-Higgs doublet solution to the LSND, MiniBooNE and muon g-2 anomalies. Phys. Rev. D **104**(5), 055028 (2021)
64. C.-H. Vincent Chang, C.-R. Chen, S.-Y. Ho, S.-Y. Tseng, Explaining the MiniBooNE anomalous excess via a leptophilic ALP-sterile neutrino coupling. Phys. Rev. D **104**(1), 015030 (2021)

65. S.N. Gninenko, The MiniBooNE anomaly and heavy neutrino decay. Phys. Rev. Lett. **103**, 241802 (2009)
66. M. Lindner, F.S. Queiroz, W. Rodejohann, X.-J. Xu, Neutrino-electron scattering: general constraints on Z' and dark photon models. J. High Energy Phys. **2018**(5), 1–29 (2018)
67. C.A. Argüelles, N. Foppiani, M. Hostert, Efficiently exploring multi-dimensional parameter spaces beyond the standard model (2022). arXiv:2205.12273
68. A. Keshavarzi et al., Muon g-2 and connection. Phys. Rev. D **102**(3), 033002 (2020)
69. Pierre Fayet, U-boson production in e+ e− annihilations, and decays, and light dark matter. Phys. Rev. D **75**(11), 115017 (2007)
70. Maxim Pospelov, Secluded U (1) below the weak scale. Phys. Rev. D **80**(9), 095002 (2009)

Chapter 10
Flavor Structure of Quark and Lepton in Modular Symmetry

Morimitsu Tanimoto

Abstract We briefly review the modular invariant flavor model. The quark and lepton mass matrices are given in terms of modular forms which are holomorphic functions of a complex parameter, the modulus τ. We present typical modular A_4 invariant models of quarks and leptons. In a model of leptons, both CP and modular symmetries are broken spontaneously by the vacuum expectation value of the modulus τ. The source of the CP violation is a non-trivial value of Re$[\tau]$ while other parameters of the model are real. The allowed region of τ is in very narrow one close to the fixed point $\tau = i$.

10.1 Introduction

Non-Abelian discrete symmetries of flavors have been developed in the last twenty years [1–11]. Among them, the A_4 flavor model is an attractive one because the A_4 group is the minimal one including a triplet irreducible representation, which allows for a natural explanation of the existence of three families of quarks and leptons [12–18]. However, it is difficult to obtain clear predictions of the A_4 flavor symmetry because of a lot of free parameters associated with scalar flavon fields.

A new approach to the quark and lepton flavors has been put forward based on the invariance under the modular transformation [19], where the model of the finite modular group $\Gamma_3 \simeq A_4$ has been presented. This work inspired further studies of the modular invariant models of the lepton flavors. It should be emphasized that there is a significant difference between the models based on the A_4 modular symmetry and those based on the traditional A_4 flavor symmetry. Yukawa couplings transform non-trivially under the modular group and are written in terms of modular forms which are holomorphic functions of a complex parameter, the modulus τ.

M. Tanimoto (✉)
Department of Physics, Niigata University, Niigata 950-2181, Japan
e-mail: tanimoto@muse.sc.niigata-u.ac.jp

G. Ricciardi et al. (eds.), *8th Workshop on Theory, Phenomenology and Experiments in Flavour Physics*, Springer Proceedings in Physics 292,
https://doi.org/10.1007/978-3-031-30459-0_10

The finite groups S_3, A_4, S_4, and A_5 are realized in modular groups [20]. Modular invariant flavor models have been also proposed on the $\Gamma_2 \simeq S_3$ [21], $\Gamma_4 \simeq S_4$ [22] and $\Gamma_5 \simeq A_5$ [23]. Phenomenological discussions of the neutrino flavor mixing have been done based on A$_4$ [24–29], S$_4$ [30–32] and A$_5$ [33]. A clear prediction of the neutrino mixing angles and the CP violating phase was given in the simple lepton mass matrices with the A$_4$ modular symmetry [25]. The quark mass matrix has been also discussed in the S_3 and A_4 modular symmetries [34–37].

The CP symmetry has been discussed in the modular invariant flavor model with the generalized CP symmetry [38–42] (see a review of Ref. [43]). Indeed, a successful model was presented in S_4 modular symmetry [44]. The moduli stabilization is also discussed in context of the CP symmetry [45, 46].

In this talk, we present successful A_4 modular invariant models of quarks and leptons.

10.2 Modular Symmetry

The modular symmetry is a geometrical symmetry of the two-dimensional torus, T^2. The two-dimensional torus is constructed as division of the two-dimensional Euclidean space R^2 by a lattice Λ, $T^2 = R^2/\Lambda$. Instead of R^2, one can use the one-dimensional complex plane. The lattice is spanned by two basis vectors, e_1 and e_2 as $m_1 e_1 + m_2 e_2$, where m_1 and m_2 are integer. The ratio,

$$\tau = \frac{e_2}{e_1}, \tag{10.1}$$

in the complex plane, represents the shape of T^2, and the parameter τ is called the modulus.

The same lattice can be spanned by other basis vectors such as

$$\begin{pmatrix} e_2' \\ e_1' \end{pmatrix} = \begin{pmatrix} a & b \\ c & d \end{pmatrix} \begin{pmatrix} e_2 \\ e_1 \end{pmatrix}, \tag{10.2}$$

where a, b, c, d are integer satisfying $ad - bc = 1$. That is the $SL(2, Z)$.

Under the above transformation, the modulus τ transforms as follows,

$$\tau \longrightarrow \tau' = \gamma\tau = \frac{a\tau + b}{c\tau + d}. \tag{10.3}$$

That is the modular symmetry [47–50]. For the element $-e$ in $SL(2, Z)$,

$$-e = \begin{pmatrix} -1 & 0 \\ 0 & -1 \end{pmatrix}, \tag{10.4}$$

the modulus τ is invariant, $\tau \longrightarrow \tau' = (-\tau)/(-1) = \tau$. Thus, the modular group is $\bar{\Gamma} = PSL(2, Z) = SL(2, Z)/\{e, -e\}$. It is sometimes called the inhomogeneous modular group. On the other hand, the group, $\Gamma = SL(2, Z)$ is called the homogeneous modular group or the full modular group.

The generators of $\Gamma \simeq SL(2, Z)$ are written by S and T,

$$S = \begin{pmatrix} 0 & 1 \\ -1 & 0 \end{pmatrix}, \qquad T = \begin{pmatrix} 1 & 1 \\ 0 & 1 \end{pmatrix}. \tag{10.5}$$

They satisfy the following algebraic relations,

$$S^4 = (ST)^3 = e\,. \tag{10.6}$$

Note that

$$S^2 = -e\,. \tag{10.7}$$

On $\bar{\Gamma} = PSL(2, Z)$, they satisfy

$$S^2 = (ST)^3 = e\,. \tag{10.8}$$

These relations are also confirmed explicitly by the following transformations:

$$S : \tau \longrightarrow -\frac{1}{\tau}, \qquad T : \tau \longrightarrow \tau + 1\,. \tag{10.9}$$

In addition to the above algebraic relations of $\bar{\Gamma} = PSL(2, Z)$, we can require $T^N = e$, i.e.

$$S^2 = (ST)^3 = T^N = e\,. \tag{10.10}$$

They can correspond to finite groups such as S_3, A_4, S_4, A_5 for $N = 2, 3, 4, 5$. In practice, we define the principal congruence subgroup $\Gamma(N)$ as

$$\Gamma(N) = \left\{ \begin{pmatrix} a & b \\ c & d \end{pmatrix} \in \Gamma \;\middle|\; \begin{pmatrix} a & b \\ c & d \end{pmatrix} = \begin{pmatrix} 1 & 0 \\ 0 & 1 \end{pmatrix} \pmod N \right\}. \tag{10.11}$$

It includes T^N, but not S or T. Then, we define the quotient $\Gamma_N = \bar{\Gamma}/\bar{\Gamma}(N)$, where the above algebraic relations are satisfied. It is found that Γ_N with $N = 2, 3, 4, 5$ are isomorphic to S_3, A_4, S_4, A_5, respectively [20].

We define $SL(2, Z_N)$ by

$$SL(2, Z_N) = \left\{ \begin{pmatrix} a & b \\ c & d \end{pmatrix} \middle|\, a, b, c, d \in Z_N, ad - bc = 1 \right\}, \tag{10.12}$$

where Z_N denotes integers modulo N. The group Γ_N is isomorphic to $PSL(2, Z_N) = SL(2, Z_N)/\{e, -e\}$ for $N > 2$, while Γ_2 is isomorphic to $SL(2, Z_2)$, because $e = -e$ in $SL(2, Z_2)$.

Similar to Γ_N, we can define $\Gamma'_N = SL(2, Z)/\Gamma(N)$, and it is the double cover of Γ_N. That is, the groups Γ'_N for $N = 3, 4, 5$ are isomorphic to the double covering groups of A_4, S_4, A_5, i.e. T', S'_4, A'_5, respectively, although Γ'_2 is isomorphic to S_3.

The upper half-plane of the modulus space τ is mapped onto itself. For example, Γ does not include the basis change. $(e_1, e_2) \longrightarrow (e_1, -e_2)$, i.e. $\tau \to -\tau$. In practice, we find

$$\mathrm{Im}(\gamma\tau) = |c\tau + d|^{-2}\mathrm{Im}(\tau) . \tag{10.13}$$

Thus, the modular group is represented on the upper half-plane of τ. Obviously one can map any value of τ on the upper half-plane into the region, $-\frac{1}{2} \leq \mathrm{Re}(\tau) \leq \frac{1}{2}$ by T^n. Furthermore, by the modular transformation one can map any value of τ on the upper half-plane into the following region:

$$-\frac{1}{2} \leq \mathrm{Re}(\tau) \leq \frac{1}{2}, \qquad |\tau| > 1 , \tag{10.14}$$

which is called the fundamental domain. Suppose that $\mathrm{Im}(\gamma\tau)$ is a maximum value among all of γ for a fixed value of τ. If $|\gamma\tau| < 1$, we map it by S, and we find

$$\mathrm{Im}(S\gamma\tau) = \frac{\mathrm{Im}(\gamma\tau)}{|\gamma\tau|^2} > \mathrm{Im}(\gamma\tau) . \tag{10.15}$$

That is inconsistent with the assumption that $\mathrm{Im}(\gamma\tau)$ is a maximum value among all of γ. That is, we find $|\gamma\tau| > 1$. Thus, we can map τ on the upper half-plane into the fundamental region by the modular transformation. The point $\tau = i$ is the fixed point under S because $S : i \to -\frac{1}{i} = i$, where Z_2 symmetry remains. Similarly, the point $\tau = e^{2\pi i/3}$ is the fixed point under ST, where Z_3 symmetry remains.

Modular forms $f_i(\tau)$ of weight k are the holomorphic functions of τ and transform as

$$f_i(\tau) \longrightarrow (c\tau + d)^k \rho(\gamma)_{ij} f_j(\tau) , \quad \gamma \in \bar{\Gamma} , \tag{10.16}$$

under the modular symmetry, where $\rho(\gamma)_{ij}$ is a unitary matrix under Γ_N. On the other hand, chiral superfields ψ_i (i denotes flavors) with weight $-k$ transform as [51]

$$\psi_i \longrightarrow (c\tau + d)^{-k} \rho(\gamma)_{ij} \psi_j . \tag{10.17}$$

We study global SUSY models. The superpotential which is built from matter fields and modular forms is assumed to be modular invariant, i.e., to have a vanishing modular weight. For given modular forms, this can be achieved by assigning appropriate weights to the matter superfields.

The kinetic terms are derived from a Kähler potential. The Kähler potential of chiral matter fields ψ_i with the modular weight $-k$ is given simply by

$$\frac{1}{[i(\bar{\tau}-\tau)]^k}\sum_i |\psi_i|^2\,, \tag{10.18}$$

where the superfield and its scalar component are denoted by the same letter, and $\bar{\tau}=\tau^*$ after taking vacuum expectation value (VEV) of τ. The canonical form of the kinetic terms is obtained by changing the normalization of parameters [25].

The general Kähler potential consistent with the modular symmetry possibly contains additional terms [52]. However, we consider only the simplest form of the Kähler potential.

10.3 Modular Invariant Flavor Model of Quarks

10.3.1 Modular Forms of A_4

We summarize modular forms of the finite modular group $\Gamma_3 \simeq A_4$. In the base of generators of A_4,

$$S=\frac{1}{3}\begin{pmatrix}-1 & 2 & 2\\ 2 & -1 & 2\\ 2 & 2 & -1\end{pmatrix}, \qquad T=\begin{pmatrix}1 & 0 & 0\\ 0 & \omega & 0\\ 0 & 0 & \omega^2\end{pmatrix}, \tag{10.19}$$

where $\omega=e^{i\frac{2}{3}\pi}$, the triplet modular forms are given as [19]:

$$\mathbf{Y}^{(2)}_{\mathbf{3}}=\begin{pmatrix}Y_1\\ Y_2\\ Y_3\end{pmatrix}=\begin{pmatrix}1+12q+36q^2+12q^3+\dots\\ -6q^{1/3}(1+7q+8q^2+\dots)\\ -18q^{2/3}(1+2q+5q^2+\dots)\end{pmatrix}, . \tag{10.20}$$

where $q=\exp[2\pi i\tau]$. They satisfy also the constraint [19]:

$$Y_2^2+2Y_1Y_3=0\,. \tag{10.21}$$

For weight 4, five modular forms are given as:

$$\mathbf{Y}^{(4)}_{\mathbf{1}}=Y_1^2+2Y_2Y_3\,,\quad \mathbf{Y}^{(4)}_{\mathbf{1'}}=Y_3^2+2Y_1Y_2\,,\quad \mathbf{Y}^{(4)}_{\mathbf{1''}}=Y_2^2+2Y_1Y_3=0\,,$$
$$\mathbf{Y}^{(4)}_{\mathbf{3}}=\begin{pmatrix}Y_1^{(4)}\\ Y_2^{(4)}\\ Y_3^{(4)}\end{pmatrix}=\begin{pmatrix}Y_1^2-Y_2Y_3\\ Y_3^2-Y_1Y_2\\ Y_2^2-Y_1Y_3\end{pmatrix}, \tag{10.22}$$

where $\mathbf{Y}_{\mathbf{1}''}^{(4)}$ vanishes due to the constraint of Eq. (10.21).

For weigh 6, there are seven modular forms as:

$$\mathbf{Y}_{\mathbf{1}}^{(6)} = Y_1^3 + Y_2^3 + Y_3^3 - 3Y_1Y_2Y_3 \,,$$
$$\mathbf{Y}_{\mathbf{3}}^{(6)} \equiv \begin{pmatrix} Y_1^{(6)} \\ Y_2^{(6)} \\ Y_3^{(6)} \end{pmatrix} = (Y_1^2 + 2Y_2Y_3) \begin{pmatrix} Y_1 \\ Y_2 \\ Y_3 \end{pmatrix} , \mathbf{Y}_{\mathbf{3}'}^{(6)} \equiv \begin{pmatrix} Y_1^{'(6)} \\ Y_2^{'(6)} \\ Y_3^{'(6)} \end{pmatrix} = (Y_3^2 + 2Y_1Y_2) \begin{pmatrix} Y_3 \\ Y_1 \\ Y_2 \end{pmatrix} . \tag{10.23}$$

We construct the fermion mass matrices by using these modular forms.

10.3.2 Quark Mass Matrices in the A_4 Modular Invariance

Let us consider a A_4 modular invariant flavor model for quarks [37]. There are freedoms for the assignments of irreducible representations and modular weights to quarks and Higgs doublets.

The simplest one is to assign the triplet of the A_4 group to three left-handed quarks, but three different singlets $(\mathbf{1}, \mathbf{1}'', \mathbf{1}')$ of A_4 to the three right-handed quarks, (u^c, c^c, t^c) and (d^c, s^c, b^c), respectively, where the sum of weights of the left-handed and the right-handed quarks is -2. Then, the weight 2 modular forms of Eq. (10.20) couple them. The Higgs fields are supposed to be A_4 singlets with weight 0. Then, three independent couplings appear in the superpotential of the up-type and down-type quark sectors, respectively, as follows:

$$w_u = \alpha_u u^c H_u \mathbf{Y}_{\mathbf{3}}^{(2)} Q + \beta_u c^c H_u \mathbf{Y}_{\mathbf{3}}^{(2)} Q + \gamma_u t^c H_u \mathbf{Y}_{\mathbf{3}}^{(2)} Q \,, \tag{10.24}$$
$$w_d = \alpha_d d^c H_d \mathbf{Y}_{\mathbf{3}}^{(2)} Q + \beta_d s^c H_d \mathbf{Y}_{\mathbf{3}}^{(2)} Q + \gamma_d b^c H_d \mathbf{Y}_{\mathbf{3}}^{(2)} Q \,, \tag{10.25}$$

where Q is the left-handed A_4 triplet quarks, and H_q is the Higgs doublet. The parameters α_q, β_q, γ_q are constant coefficients. Assign the A_4 triplet Q as $((d_L, u_L), (s_L, c_L), (b_L, t_L))$. By using the decomposition of the A_4 tensor product in Appendix A, the superpotentials in Eqs. (10.24) and (10.25) give the mass matrix of quarks, which is written in terms of modular forms of weight 2 in Eq. (10.20) as:

$$M_q = v_q \begin{pmatrix} \alpha_q & 0 & 0 \\ 0 & \beta_q & 0 \\ 0 & 0 & \gamma_q \end{pmatrix} \begin{pmatrix} Y_1 & Y_3 & Y_2 \\ Y_2 & Y_1 & Y_3 \\ Y_3 & Y_2 & Y_1 \end{pmatrix}_{RL} , \qquad (q = u, d) \,, \tag{10.26}$$

where the constant v_q $(q = u, d)$ is the VEV of the neutral component of the Higgs field H_q. Parameters α_q, β_q, γ_q are taken to be real without loss of generality, and they can be adjusted to the observed quark masses. The remained parameter is only the modulus τ. The numerical study of the quark mass matrix in Eq. (10.26) is rather

Table 10.1 Representations and weights $-k_I$ for MSSM fields, and modular forms

	Q	$(d^c,\ s^c,\ b^c)$	$(u^c,\ c^c,\ t^c)$	H_q	$\mathbf{Y_3^{(2)}},\ \mathbf{Y_3^{(6)}},\ \mathbf{Y_{3'}^{(6)}}$
$SU(2)$	**2**	**1**	**1**	**2**	**1**
A_4	**3**	$(\mathbf{1}, \mathbf{1}'', \mathbf{1}')$	$(\mathbf{1}, \mathbf{1}'', \mathbf{1}')$	**1**	$\mathbf{3},\quad \mathbf{3},\quad \mathbf{3}'$
$-k_I$	-2	$(0,\ 0,\ 0)$	$(-4,\ -4,\ -4)$	0	$k=2,\quad k=6,\quad k=6$

easy. However, it is impossible to reproduce observed hierarchical three CKM mixing angles by fixing one complex parameter τ.

In order to obtain realistic quark mass matrices [37], we use modular forms of weight 6 in Eq. (10.23). As a simple model, we take modular forms of weight 6 only for the up-type quark mass matrix while the down-type quark one is still given in terms of modular forms of weight 2 such as Eq. (10.26).

Then, we have six independent couplings in the superpotential of the up-quark sector as:

$$w_u = \alpha_u u^c H_u \mathbf{Y_3^{(6)}} Q + \alpha'_u u^c H_u \mathbf{Y_{3'}^{(6)}} Q + \beta_u c^c H_u \mathbf{Y_3^{(6)}} Q + \beta'_u c^c H_u \mathbf{Y_{3'}^{(6)}} Q \\ + \gamma_u t^c H_u \mathbf{Y_3^{(6)}} Q + \gamma'_u t^c H_q \mathbf{Y_{3'}^{(6)}} Q\,, \tag{10.27}$$

where assignments of representations and weights for MSSM fields are given in Table 10.1. The up-type quark mass matrix is written as:

$$M_u = v_u \begin{pmatrix} \alpha_u & 0 & 0 \\ 0 & \beta_u & 0 \\ 0 & 0 & \gamma_u \end{pmatrix} \left[\begin{pmatrix} Y_1^{(6)} & Y_3^{(6)} & Y_2^{(6)} \\ Y_2^{(6)} & Y_1^{(6)} & Y_3^{(6)} \\ Y_3^{(6)} & Y_2^{(6)} & Y_1^{(6)} \end{pmatrix} + \begin{pmatrix} g_{u1} & 0 & 0 \\ 0 & g_{u2} & 0 \\ 0 & 0 & g_{u3} \end{pmatrix} \begin{pmatrix} Y_1^{'(6)} & Y_3^{'(6)} & Y_2^{'(6)} \\ Y_2^{'(6)} & Y_1^{'(6)} & Y_3^{'(6)} \\ Y_3^{'(6)} & Y_2^{'(6)} & Y_1^{'(6)} \end{pmatrix} \right], \tag{10.28}$$

where $g_{u1} = \alpha'_u/\alpha_u$, $g_{u2} = \beta'_u/\beta_u$ and $g_{u3} = \gamma'_u/\gamma_u$ are complex parameters while α_u, β_u and γ_u are real. On the other hand, the down-type quark mass matrix is given as:

$$M_d = v_d \begin{pmatrix} \alpha_d & 0 & 0 \\ 0 & \beta_d & 0 \\ 0 & 0 & \gamma_d \end{pmatrix} \begin{pmatrix} Y_1 & Y_3 & Y_2 \\ Y_2 & Y_1 & Y_3 \\ Y_3 & Y_2 & Y_1 \end{pmatrix}, \tag{10.29}$$

where the chiralities of the mass matrix, L and R are taken as $[M_{u(d)}]_{LR}$. We will fix the modulus τ phenomenologically by using quark mass matrices in Eqs. (10.28) and (10.29).

10.3.3 Fixing τ by Observed CKM

In order to obtain the left-handed flavor mixing, we calculate $M_d^\dagger M_d$ and $M_u^\dagger M_u$. At first, we take a random point of τ and g_{ui} which are scanned in the complex plane by generating random numbers. The modulus τ is scanned in the fundamental domain of the modular symmetry.

We input quark masses in order to constrain model parameters. Since the modulus τ obtains the expectation value by the breaking of the modular invariance at the high mass scale, the quark masses are put at the GUT scale. The observed masses and CKM parameters run to the GUT scale by the renormalization group equations (RGEs). In our work, we adopt numerical values of Yukawa couplings of quarks at the GUT scale 2×10^{16} GeV with $\tan\beta = 5$ in the framework of the minimal SUSY breaking scenarios [53, 54]:

$$\begin{aligned}
&y_d = (4.81 \pm 1.06) \times 10^{-6},\ y_s = (9.52 \pm 1.03) \times 10^{-5},\ y_b = (6.95 \pm 0.175)\times 10^{-3},\\
&y_u = (2.92 \pm 1.81) \times 10^{-6},\ y_c = (1.43 \pm 0.100) \times 10^{-3},\ y_t = 0.534 \pm 0.0341,
\end{aligned} \tag{10.30}$$

which give quark masses as $m_q = y_q v_H$ with $v_H = 174$ GeV.

We also use the following CKM mixing angles to focus on parameter regions consistent with the experimental data at the GUT scale 2×10^{16} GeV, where $\tan\beta = 5$ is taken [53, 54]:

$$\theta_{12}^{\rm CKM} = 13.027^\circ \pm 0.0814^\circ,\ \theta_{23}^{\rm CKM} = 2.054^\circ \pm 0.384^\circ,\ \theta_{13}^{\rm CKM} = 0.1802^\circ \pm 0.0281^\circ. \tag{10.31}$$

Here $\theta_{ij}^{\rm CKM}$ is given in the PDG notation of the CKM matrix $V_{\rm CKM}$ [55]. The CP violating phase is also given as:

$$\delta_{CP} = 69.21^\circ \pm 6.19^\circ\ , \tag{10.32}$$

in the PDG notation. The errors in Eqs. (10.30), (10.31) and (10.32) represent 1σ interval. The CKM elements V_{ij} at the GUT scale 2×10^{16} GeV are given by using these angles and the phase.

In our model, we have four complex parameters, τ, g_{u1}, g_{u2} and g_{u3} after inputting six quark masses. These eight real parameters are scanned to reproduce the observed three CKM mixing angles and the CP violating phase with three times 1σ error interval in Eqs. (10.31) and (10.32).

We have succeeded to reproduce completely four observed CKM elements in the parameter ranges of Table 10.2. The modulus τ is close to i, which is the fixed point of the modular symmetry.

In Table 10.3, we present one parameter set and calculated CKM elements, which is the best-fit point, that is, its χ^2 is minimum. The magnitudes of g_{qi} are at most of order $\mathcal{O}(0.5)$. Ratios of α_q/γ_q and β_q/γ_q $(q = u, d)$ correspond to the observed quark mass hierarchy.

Table 10.2 Parameter ranges consistent with CKM mixing angles and CP phase δ_{CP}

$\lvert\text{Re}[\tau]\rvert$	$\text{Im}[\tau]$	$\lvert g_{u1}\rvert$	$\text{Arg}\, g_{u1}$	$\lvert g_{u2}\rvert$	$\text{Arg}\, g_{u2}$	$\lvert g_{u3}\rvert$	$\text{Arg}\, g_{u3}$
[0, 0.09]	[0.99, 1.09]	[0.01, 0.86]	$[-\pi, \pi]$	[0.14, 1.29]	[−2.3, 1.6]	[0.02, 0.07]	$[-\pi, \pi]$

Table 10.3 Values of parameters and output of CKM parameters at the best-fit point

τ	$-0.038 + 1.05\,i$	$\lvert V_{us}\rvert$	0.225
g_{u1}	$-0.147 + 0.118\,i$	$\lvert V_{cb}\rvert$	0.029
g_{u2}	$-0.091 - 0.425\,i$	$\lvert V_{ub}\rvert$	0.0030
g_{u3}	$0.027 + 0.0197\,i$	δ_{CP}	76.9°
α_u/γ_u	4.33×10^{-5}	χ^2	0.46
β_u/γ_u	3.85×10^{-3}		
α_d/γ_d	1.45×10^{-2}		
β_d/γ_d	4.26×10^{-3}		

In conclusion, our quark mass matrices with the A_4 modular symmetry reproduce the observed CKM mixing matrix very well at nearby fixed point $\tau = i$.

10.4 Spontaneous CP Violation in A_4 Model of Leptons

10.4.1 CP Transformation in Flavor Space

The modular symmetry meets the CP symmetry because both are the discrete symmetry. Let us summarize the CP transformations of chiral superfields and modular multiplets as follows:

$$\tau \xrightarrow{\text{CP}} -\tau^*, \quad \psi(x) \xrightarrow{\text{CP}} \mathbf{X_r}\overline{\psi}(x_P), \quad \mathbf{Y_r^{(k)}}(\tau) \xrightarrow{\text{CP}} \mathbf{Y_r^{(k)}}(-\tau^*) = \mathbf{X_r}\mathbf{Y_r^{(k)*}}(\tau), \tag{10.33}$$

where $\mathbf{X_r}$ satisfies [56]

$$\mathbf{X_r}\rho_{\mathbf{r}}^*(g)\mathbf{X_r}^{-1} = \rho_{\mathbf{r}}(g'), \qquad g, g' \in G. \tag{10.34}$$

One can take $\mathbf{X_r} = \mathbf{1_r}$ in a basis of symmetric generators of S and T.

10.4.2 A_4 Model of CP Violation by Modulus τ

There are the modular A_4 invariant models with the generalized CP symmetry [57, 58]. Both CP and modular symmetries are broken spontaneously by VEV of the modulus τ. We discuss the phenomenological implication of a simple model [57], that is the neutrino mixing angles and the CP violating Dirac phase of leptons, which is expected to be observed in the future.

We assign the A_4 representation and weight for superfields of leptons in Table 10.4, where the three left-handed lepton doublets compose a A_4 triplet L, and the right-handed charged leptons e^c, μ^c and τ^c are A_4 singlets. The weights of the superfields of left-handed leptons and right-handed charged leptons are -2 and 0, respectively. Then, the simple lepton mass matrices for charged leptons and neutrinos are obtained [57].

The superpotential of the charged lepton mass term is given in terms of modular forms of weight 2, $\mathbf{Y}_3^{(2)}$. It is given as:

$$w_E = \alpha_e e^c H_d \mathbf{Y}_3^{(2)} L + \beta_e \mu^c H_d \mathbf{Y}_3^{(2)} L + \gamma_e \tau^c H_d \mathbf{Y}_3^{(2)} L\,, \tag{10.35}$$

where L is the left-handed A_4 triplet leptons. We can take real for α_e, β_e and γ_e. Under CP, the superfields transform as:

$$e^c \xrightarrow{CP} \mathbf{X_1}^* \overline{e}^c\,, \quad \mu^c \xrightarrow{CP} \mathbf{X_{1''}}^* \overline{\mu}^c\,, \quad \tau^c \xrightarrow{CP} \mathbf{X_{1'}}^* \overline{\tau}^c\,, \quad L \xrightarrow{CP} \mathbf{X_3} \overline{L}\,, \tag{10.36}$$

and $H_d \xrightarrow{CP} \overline{H}_d$. In the symmetric base of S and T as seen in Eq. (10.19) of Appendix A, we can choose $X_3 = 1_3$ and $\mathbf{X_1} = \mathbf{X_{1'}} = \mathbf{X_{1''}} = 1$.

Taking (e_L, μ_L, τ_L) in the flavor base, the charged lepton mass matrix M_E is simply written as:

$$M_E(\tau) = v_d \begin{pmatrix} \alpha_e & 0 & 0 \\ 0 & \beta_e & 0 \\ 0 & 0 & \gamma_e \end{pmatrix} \begin{pmatrix} Y_1(\tau) & Y_3(\tau) & Y_2(\tau) \\ Y_2(\tau) & Y_1(\tau) & Y_3(\tau) \\ Y_3(\tau) & Y_2(\tau) & Y_1(\tau) \end{pmatrix}, \tag{10.37}$$

where v_d is VEV of the neutral component of H_d, and coefficients α_e, β_e and γ_e are taken to be real without loss of generality. Under CP transformation, the mass matrix M_E is transformed following from (10.33) as:

Table 10.4 Representations and weights k for MSSM fields, and modular forms

	L	(e^c, μ^c, τ^c)	H_u	H_d	$\mathbf{Y_r}^{(2)}$, $\mathbf{Y_r}^{(4)}$
$SU(2)$	**2**	**1**	**2**	**2**	**1**
A_4	**3**	$(\mathbf{1}, \mathbf{1''}, \mathbf{1'})$	**1**	**1**	**3**, $\{\mathbf{3}, \mathbf{1}, \mathbf{1'}\}$
$-k_I$	-2	(0, 0, 0)	0	0	$k = 2$, 4

$$M_E(\tau) \xrightarrow{CP} M_E(-\tau^*) = M_E^*(\tau) = v_d \begin{pmatrix} \alpha_e & 0 & 0 \\ 0 & \beta_e & 0 \\ 0 & 0 & \gamma_e \end{pmatrix} \begin{pmatrix} Y_1(\tau)^* & Y_3(\tau)^* & Y_2(\tau)^* \\ Y_2(\tau)^* & Y_1(\tau)^* & Y_3(\tau)^* \\ Y_3(\tau)^* & Y_2(\tau)^* & Y_1(\tau)^* \end{pmatrix}. \tag{10.38}$$

Let us discuss the neutrino mass matrix. Suppose neutrinos to be Majorana particles. By using the Weinberg operator, the superpotential of the neutrino mass term, w_ν is given as:

$$w_\nu = -\frac{1}{\Lambda}(H_u H_u L L \mathbf{Y}_{\mathbf{r}}^{(4)})_{\mathbf{1}}\,, \tag{10.39}$$

where Λ is a relevant cutoff scale. Since the left-handed lepton doublet has weight -2, the superpotential is given in terms of modular forms of weight 4, $\mathbf{Y}_{\mathbf{1}}^{(4)}$, $\mathbf{Y}_{\mathbf{1}'}^{(4)}$, $\mathbf{Y}_{\mathbf{3}}^{(4)}$, which are given in Eq. (10.22).

By putting v_u for VEV of the neutral component of H_u and using the tensor products of A_4 in Appendix A, we have

$$\begin{aligned} w_\nu = &\frac{v_u^2}{\Lambda}\left[\begin{pmatrix} 2\nu_e\nu_e - \nu_\mu\nu_\tau - \nu_\tau\nu_\mu \\ 2\nu_\tau\nu_\tau - \nu_e\nu_\mu - \nu_\mu\nu_\tau \\ 2\nu_\mu\nu_\mu - \nu_\tau\nu_e - \nu_e\nu_\tau \end{pmatrix} \otimes \mathbf{Y}_{\mathbf{3}}^{(4)} \right.\\ &\left. + (\nu_e\nu_e + \nu_\mu\nu_\tau + \nu_\tau\nu_\mu) \otimes g_1^\nu \mathbf{Y}_{\mathbf{1}}^{(4)} + (\nu_e\nu_\tau + \nu_\mu\nu_\mu + \nu_\tau\nu_e) \otimes g_2^\nu \mathbf{Y}_{\mathbf{1}'}^{(4)} \right] \\ = &\frac{v_u^2}{\Lambda}\left[(2\nu_e\nu_e - \nu_\mu\nu_\tau - \nu_\tau\nu_\mu)Y_1^{(4)} + (2\nu_\tau\nu_\tau - \nu_e\nu_\mu - \nu_\mu\nu_e)Y_3^{(4)}\right.\\ &+(2\nu_\mu\nu_\mu - \nu_\tau\nu_e - \nu_e\nu_\tau)Y_2^{(4)} \\ &\left.+ (\nu_e\nu_e + \nu_\mu\nu_\tau + \nu_\tau\nu_\mu) g_1^\nu \mathbf{Y}_{\mathbf{1}}^{(4)} + (\nu_e\nu_\tau + \nu_\mu\nu_\mu + \nu_\tau\nu_e) g_2^\nu \mathbf{Y}_{\mathbf{1}'}^{(4)}\right], \end{aligned} \tag{10.40}$$

where g_1^ν, g_2^ν are complex parameters in general. The neutrino mass matrix is written as follows:

$$\begin{aligned} M_\nu(\tau) = \frac{v_u^2}{\Lambda}&\left[\begin{pmatrix} 2Y_1^{(4)}(\tau) & -Y_3^{(4)}(\tau) & -Y_2^{(4)}(\tau) \\ -Y_3^{(4)}(\tau) & 2Y_2^{(4)}(\tau) & -Y_1^{(4)}(\tau) \\ -Y_2^{(4)}(\tau) & -Y_1^{(4)}(\tau) & 2Y_3^{(4)}(\tau) \end{pmatrix} + g_1^\nu \mathbf{Y}_{\mathbf{1}}^{(4)}(\tau) \begin{pmatrix} 1 & 0 & 0 \\ 0 & 0 & 1 \\ 0 & 1 & 0 \end{pmatrix}\right. \\ &\left. + g_2^\nu \mathbf{Y}_{\mathbf{1}'}^{(4)}(\tau) \begin{pmatrix} 0 & 0 & 1 \\ 0 & 1 & 0 \\ 1 & 0 & 0 \end{pmatrix}\right]. \end{aligned} \tag{10.41}$$

Under CP transformation, the mass matrix M_ν is transformed following from (10.33) as:

$$M_\nu(\tau) \xrightarrow{CP} M_\nu(-\tau^*) = M_\nu^*(\tau)$$

$$= \frac{v_u^2}{\Lambda}\left[\begin{pmatrix} 2Y_1^{(4)*}(\tau) & -Y_3^{(4)*}(\tau) & -Y_2^{(4)*}(\tau) \\ -Y_3^{(4)*}(\tau) & 2Y_2^{(4)*}(\tau) & -Y_1^{(4)*}(\tau) \\ -Y_2^{(4)*}(\tau) & -Y_1^{(4)*}(\tau) & 2Y_3^{(4)*}(\tau) \end{pmatrix} + g_1^{\nu*}\mathbf{Y}_{\mathbf{1}}^{(4)*}(\tau)\begin{pmatrix} 1 & 0 & 0 \\ 0 & 0 & 1 \\ 0 & 1 & 0 \end{pmatrix}\right.$$

$$\left. + g_2^{\nu*}\mathbf{Y}_{\mathbf{1}'}^{(4)*}(\tau)\begin{pmatrix} 0 & 0 & 1 \\ 0 & 1 & 0 \\ 1 & 0 & 0 \end{pmatrix}\right] . \tag{10.42}$$

In a CP conserving modular invariant theory, both CP and modular symmetries are broken spontaneously by VEV of the modulus τ. However, there exist certain values of τ which conserve CP while breaking the modular symmetry. Obviously, this is the case if τ is left invariant by CP, i.e.

$$\tau \xrightarrow{CP} -\tau^* = \tau \ , \tag{10.43}$$

which indicates τ lies on the imaginary axis, $\mathrm{Re}[\tau] = 0$. In addition to $\mathrm{Re}[\tau] = 0$, CP is conserved at the boundary of the fundamental domain. Then, one has

$$M_E(\tau) = M_E^*(\tau) , \qquad M_\nu(\tau) = M_\nu^*(\tau) , \tag{10.44}$$

which leads to g_1^ν and g_2^ν being real. Since parameters α_e, β_e, γ_e are also real, the source of the CP violation is only non-trivial $\mathrm{Re}[\tau]$ after breaking the modular symmetry.

10.4.3 Numerical Result

In this model, VEV of τ breaks the CP symmetry as well as the modular invariance. The source of the CP violation is only the real part of τ. It is interesting to ask whether the spontaneous CP violation is realized due to the value of τ, which is consistent with observed lepton mixing angles and neutrino masses. If this is the case, the CP violating Dirac phase and Majorana phases are predicted clearly under the fixed value of τ.

The framework of our calculations is given as follows. Parameter ratios α_e/γ_e and β_e/γ_e are given in terms of charged lepton masses and τ. Therefore, the lepton mixing angles, the Dirac phase and Majorana phases are given by our model parameters g_1^ν and g_2^ν in addition to the value of τ.

Taking $\tan\beta = 5$ as a bench mark, the input charged lepton masses are given at the GUT scale 2×10^{16} GeV [53, 54]:

$$y_e = (1.97 \pm 0.024) \times 10^{-6}, \qquad y_\mu = (4.16 \pm 0.050) \times 10^{-4},$$
$$y_\tau = (7.07 \pm 0.073) \times 10^{-3}, \tag{10.45}$$

where lepton masses are given by $m_\ell = y_\ell v_H$ with $v_H = 174$ GeV. We also input the lepton mixing angles and neutrino mass parameters which are given by NuFit 5.0 [59]. In our analysis, the Dirac CP phase δ is output because its observed range is too wide at $3\,\sigma$ confidence level. We investigate two possible cases of neutrino masses m_i, which are the normal hierarchy (NH), $m_3 > m_2 > m_1$, and the inverted hierarchy (IH), $m_2 > m_1 > m_3$. Neutrino masses and mixings are obtained by diagonalizing $M_E^\dagger M_E$ and $M_\nu^\dagger M_\nu$. We also investigate the effective mass for the $0\nu\beta\beta$ decay, $\langle m_{ee}\rangle$ and the sum of three neutrino masses $\sum m_i$ since it is constrained by the recent cosmological data, which is the upper-bound $\sum m_i \leq 120\,\text{meV}$ obtained at the 95% confidence level [60, 61].

Let us discuss numerical results for NH of neutrino masses. Since the spontaneous CP violation in Type IIB string theory is possibly realized at nearby fixed points, where the moduli stabilization is performed in a controlled way [62, 63]. There are two fixed points in the fundamental domain of $PSL(2, Z)$, $\tau = i$ and $\tau = \omega$. Indeed, the viable τ of our lepton mass matrices is found around $\tau = i$. We scan τ around i while neutrino couplings g_1^ν and g_2^ν are scanned in the real space of $[-10, 10]$. As a measure of good-fit, we adopt the sum of one-dimensional χ^2 function for four accurately known dimensionless observables $\Delta m^2_{\text{atm}}/\Delta m^2_{\text{sol}}$, $\sin^2\theta_{12}$, $\sin^2\theta_{23}$ and $\sin^2\theta_{13}$ in NuFit 5.0 [59]. In addition, we employ Gaussian approximations for fitting m_e/m_τ and m_μ/m_τ by using the data of PDG [55].

In Fig. 10.1 we show the allowed region on the τ plane, where three mixing angles and $\Delta m^2_{\text{atm}}/\Delta m^2_{\text{sol}}$ are consistent with observed ones. The Dark, middle and light region correspond to 2σ, 3σ and 5σ confidence levels, respectively.

The predicted range of τ is close to the fixed point $\tau = i$. Due to restricted Re $[\tau]$, the CP violating Dirac phase δ_{CP}, which is defined in PDG, is predicted clearly. In Fig. 10.2, we show prediction of δ_{CP} versus the sum of neutrino masses $\sum m_i$. It is remarked that δ_{CP} is almost independent of $\sum m_i$. The predicted ranges of δ_{CP} are narrow such as $[98°, 110°]$ and $[250°, 262°]$ at $3\,\sigma$ confidence level (yellow). The predicted ranges $[98°, 110°]$ and $[250°, 262°]$ correspond to Re $[\tau] = (0.073–0.083)$ and Re $[\tau] = -(0.073–0.083)$, respectively. The predicted $\sum m_i$ is in $[82, 102]$ meV

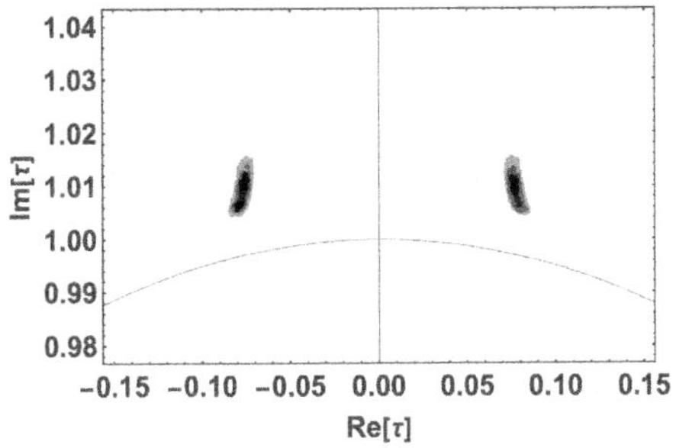

Fig. 10.1 The dark, middle and light regions correspond to 2σ, 3σ, 5σ confidence levels, respectively. The solid curve is the boundary of the fundamental domain, $|\tau| = 1$

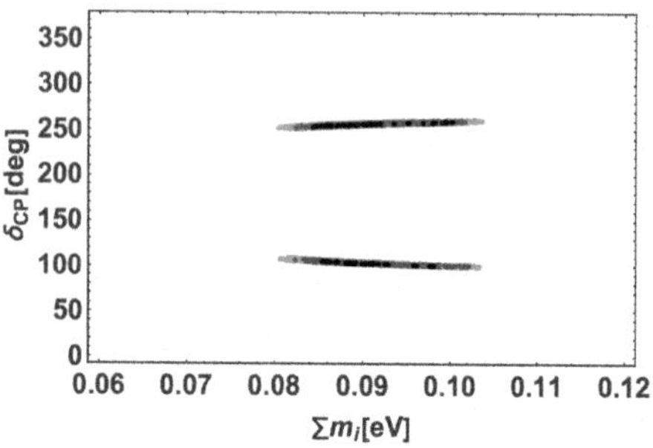

Fig. 10.2 For NH. The dark, middle and light regions denote same ones in Fig. 10.1. The vertical line denotes the observed upperbound of the sum of neutrino masses

for $3\,\sigma$ confidence level (middle). The minimal cosmological model, CDM + $\sum m_i$, provides the upper-bound $\sum m_i < 120\,\text{meV}$ [60, 61]. Thus, our predicted sum of neutrino masses is consistent with the cosmological bound 120 meV.

We can calculate the effective mass $\langle m_{ee}\rangle$ for the $0\nu\beta\beta$ decay. The predicted $\langle m_{ee}\rangle$ is in $[12.5,\ 20.5]\,\text{meV}$ for $3\,\sigma$ confidence level. The prediction of $\langle m_{ee}\rangle \simeq 20\,\text{meV}$ will be testable in the future experiments of the neutrinoless double beta decay.

For the case of inverted hierarchy (IH) of neutrino masses, there are no allowed regions of $2\,\sigma$ and $3\,\sigma$ confidence levels, but only $5\,\sigma$ confidence level. Moreover, the sum of neutrino masses are in the range of $[134,\ 180]\,\text{meV}$. Therefore, the IH of neutrino masses are unfavored in this model.

In our numerical calculations, we have not included the effects of the renormalization group equation in the lepton mixing angles and neutrino mass ratio $\Delta m^2_{\text{sol}}/\Delta m^2_{\text{atm}}$. Those corrections are very small between the electroweak and GUT scales in the case of $\tan\beta \leq 5$ unless neutrino masses are almost degenerate.

10.5 Summary and Discussions

We have presented typical quark and lepton mass matrices in the A_4 modular symmetry. If flavors of quarks and leptons are originated from a same two-dimensional compact space, quarks and leptons have the same flavor symmetry and the same value of the modulus τ.

We have presented the viable model for quark mass matrices, in which the down-type quark mass matrix is constructed by modular forms of weight 2 while the up-type quark mass matrix is constructed by modular forms of weight 6. The observed CKM parameters fix the value of τ, which is close to the fixed point $\tau = i$.

The modular invariant A_4 model of lepton flavors is constructed combining with the generalized CP symmetry. In our model, both CP and modular symmetries are broken spontaneously by VEV of the modulus τ. The source of the CP violation is a non-trivial value of Re[τ] while parameters of neutrinos g_1^ν and g_2^ν are real.

As well as in the quark mass matrices, we have found allowed region of τ close to the fixed point $\tau = i$, which is consistent with the observed lepton mixing angles and lepton masses for NH at $2\,\sigma$ confidence level. The CP violating Dirac phase δ_{CP} is predicted clearly in $[98^\circ, 110^\circ]$ and $[250^\circ, 262^\circ]$ at $3\,\sigma$ confidence level. The predicted $\sum m_i$ is in $[82,\ 102]$ meV with $3\,\sigma$ confidence level. The case of IH of neutrino masses is unfavored because the sum of neutrino masses is predicted in $\sum m_i = [134,\ 180]$ meV.

In our work, we have obtained τ by observables of quarks and leptons phenomenologically. On the other hand, one also should pay attention to the recent theoretical work of the moduli stabilization from the viewpoint of modular flavor symmetries [45, 46, 62–64]. The study of modulus τ is interesting to reveal the flavor theory in both theoretical and phenomenological aspects.

Acknowledgements The authors would like to thank H. Okada and T. Kobayashi for important collaborations.

Appendix A: Tensor Product of A_4 Group

We take the generators of A_4 group for the triplet as follows:

$$S = \frac{1}{3}\begin{pmatrix} -1 & 2 & 2 \\ 2 & -1 & 2 \\ 2 & 2 & -1 \end{pmatrix}, \qquad T = \begin{pmatrix} 1 & 0 & 0 \\ 0 & \omega & 0 \\ 0 & 0 & \omega^2 \end{pmatrix}, \tag{10.46}$$

where $\omega = e^{i\frac{2}{3}\pi}$ for a triplet. In this base, the multiplication rule is

$$\begin{aligned} \begin{pmatrix} a_1 \\ a_2 \\ a_3 \end{pmatrix}_{\mathbf{3}} \otimes \begin{pmatrix} b_1 \\ b_2 \\ b_3 \end{pmatrix}_{\mathbf{3}} &= (a_1 b_1 + a_2 b_3 + a_3 b_2)_{\mathbf{1}} \oplus (a_3 b_3 + a_1 b_2 + a_2 b_1)_{\mathbf{1}'} \\ &\oplus (a_2 b_2 + a_1 b_3 + a_3 b_1)_{\mathbf{1}''} \\ &\oplus \frac{1}{3}\begin{pmatrix} 2a_1 b_1 - a_2 b_3 - a_3 b_2 \\ 2a_3 b_3 - a_1 b_2 - a_2 b_1 \\ 2a_2 b_2 - a_1 b_3 - a_3 b_1 \end{pmatrix}_{\mathbf{3}} \oplus \frac{1}{2}\begin{pmatrix} a_2 b_3 - a_3 b_2 \\ a_1 b_2 - a_2 b_1 \\ a_3 b_1 - a_1 b_3 \end{pmatrix}_{\mathbf{3}}, \end{aligned}$$

$$\mathbf{1} \otimes \mathbf{1} = \mathbf{1}\,, \qquad \mathbf{1}' \otimes \mathbf{1}' = \mathbf{1}''\,, \qquad \mathbf{1}'' \otimes \mathbf{1}'' = \mathbf{1}'\,, \qquad \mathbf{1}' \otimes \mathbf{1}'' = \mathbf{1}\,, \tag{10.47}$$

where

$$T(\mathbf{1}') = \omega\,, \qquad T(\mathbf{1}'') = \omega^2. \tag{10.48}$$

More details are shown in the review [2, 3].

References

1. G. Altarelli, F. Feruglio, Rev. Mod. Phys. **82**, 2701 (2010). arXiv:1002.0211 [hep-ph]
2. H. Ishimori, T. Kobayashi, H. Ohki, Y. Shimizu, H. Okada, M. Tanimoto, Prog. Theor. Phys. Suppl. **183**, 1 (2010). arXiv:1003.3552 [hep-th]
3. H. Ishimori, T. Kobayashi, H. Ohki, H. Okada, Y. Shimizu, M. Tanimoto, Lect. Notes Phys. **858**, 1 (2012). Springer
4. T. Kobayashi, H. Ohki, H. Okada, Y. Shimizu, M. Tanimoto, Lect. Notes Phys. **995**, 1 (2022). Springer. https://doi.org/10.1007/978-3-662-64679-3
5. D. Hernandez, A.Y. Smirnov, Phys. Rev. D **86**, 053014 (2012). arXiv:1204.0445 [hep-ph]
6. S.F. King, C. Luhn, Rept. Prog. Phys. **76**, 056201 (2013). arXiv:1301.1340 [hep-ph]
7. S.F. King, A. Merle, S. Morisi, Y. Shimizu, M. Tanimoto, New J. Phys. **16**, 045018 (2014). arXiv:1402.4271 [hep-ph]
8. M. Tanimoto, AIP Conf. Proc. **1666**, 120002 (2015)
9. S.F. King, Prog. Part. Nucl. Phys. **94**, 217 (2017). arXiv:1701.04413 [hep-ph]
10. S.T. Petcov, Eur. Phys. J. C **78**(9), 709 (2018). arXiv:1711.10806 [hep-ph]
11. F. Feruglio, A. Romanino. arXiv:1912.06028 [hep-ph]
12. E. Ma, G. Rajasekaran, Phys. Rev. D **64**, 113012 (2001). arXiv:hep-ph/0106291
13. K.S. Babu, E. Ma, J.W.F. Valle, Phys. Lett. B **552**, 207 (2003). arXiv:hep-ph/0206292
14. G. Altarelli, F. Feruglio, Nucl. Phys. B **720**, 64 (2005). [hep-ph/0504165]
15. G. Altarelli, F. Feruglio, Nucl. Phys. B **741**, 215 (2006). [hep-ph/0512103]
16. Y. Shimizu, M. Tanimoto, A. Watanabe, Prog. Theor. Phys. **126**, 81 (2011). arXiv:1105.2929 [hep-ph]
17. S.T. Petcov, A.V. Titov, Phys. Rev. D **97**(11), 115045 (2018). arXiv:1804.00182 [hep-ph]
18. S.K. Kang, Y. Shimizu, K. Takagi, S. Takahashi, M. Tanimoto, PTEP **2018**(8), 083B01 (2018). arXiv:1804.10468 [hep-ph]
19. F. Feruglio, in From My Vast Repertoire ...: Guido Altarelli's Legacy, ed. by A. Levy, S. Forte, G. Ridolfi (2019), pp.227–266. arXiv:1706.08749 [hep-ph]
20. R. de Adelhart Toorop, F. Feruglio, C. Hagedorn, Nucl. Phys. B **858**, 437 (2012). arXiv:1112.1340 [hep-ph]
21. T. Kobayashi, K. Tanaka, T.H. Tatsuishi, Phys. Rev. D **98**(1), 016004 (2018). arXiv:1803.10391 [hep-ph]
22. J.T. Penedo, S.T. Petcov, Nucl. Phys. B **939**, 292 (2019). arXiv:1806.11040 [hep-ph]
23. P.P. Novichkov, J.T. Penedo, S.T. Petcov, A.V. Titov, JHEP **1904**, 174 (2019). arXiv:1812.02158 [hep-ph]
24. J.C. Criado, F. Feruglio, SciPost Phys. **5**(5), 042 (2018). arXiv:1807.01125 [hep-ph]
25. T. Kobayashi, N. Omoto, Y. Shimizu, K. Takagi, M. Tanimoto, T.H. Tatsuishi, JHEP **1811**, 196 (2018). arXiv:1808.03012 [hep-ph]
26. G.J. Ding, S.F. King, X.G. Liu, JHEP **1909**, 074 (2019). arXiv:1907.11714 [hep-ph]
27. D. Zhang, Nucl. Phys. B **952**, 114935 (2020). arXiv:1910.07869 [hep-ph]
28. H. Okada, M. Tanimoto. arXiv:2005.00775 [hep-ph]
29. H. Okada, M. Tanimoto. arXiv:2009.14242 [hep-ph]
30. P.P. Novichkov, J.T. Penedo, S.T. Petcov, A.V. Titov, JHEP **1904**, 005 (2019). arXiv:1811.04933 [hep-ph]
31. T. Kobayashi, Y. Shimizu, K. Takagi, M. Tanimoto, T.H. Tatsuishi, JHEP **02**, 097 (2020). arXiv:1907.09141 [hep-ph]
32. X. Wang, S. Zhou, JHEP **05**, 017 (2020). arXiv:1910.09473 [hep-ph]
33. G.J. Ding, S.F. King, X.G. Liu, Phys. Rev. D **100**(11), 115005 (2019). arXiv:1903.12588 [hep-ph]
34. T. Kobayashi, Y. Shimizu, K. Takagi, M. Tanimoto, T.H. Tatsuishi, H. Uchida, Phys. Lett. B **794**, 114 (2019). arXiv:1812.11072 [hep-ph]
35. T. Kobayashi, Y. Shimizu, K. Takagi, M. Tanimoto, T.H. Tatsuishi, PTEP **2020**(5), 053B05 (2020). arXiv:1906.10341 [hep-ph]

36. H. Okada, M. Tanimoto, Phys. Lett. B **791**, 54 (2019). arXiv:1812.09677 [hep-ph]
37. H. Okada, M. Tanimoto, Eur. Phys. J. C **81**(1), 52 (2021). arXiv:1905.13421 [hep-ph]
38. G. Ecker, W. Grimus, W. Konetschny, Nucl. Phys. B **191**, 465–492 (1981)
39. G. Ecker, W. Grimus, H. Neufeld, Nucl. Phys. B **247**, 70–82 (1984)
40. G. Ecker, W. Grimus, H. Neufeld, J. Phys. A **20**, L807 (1987)
41. H. Neufeld, W. Grimus, G. Ecker, Int. J. Mod. Phys. A **3**, 603–616 (1988)
42. W. Grimus, M.N. Rebelo, Phys. Rept. **281**, 239–308 (1997). arXiv:hep-ph/9506272 [hep-ph]
43. G.C. Branco, R.G. Felipe, F.R. Joaquim, Rev. Mod. Phys. **84**, 515 (2012). arXiv:1111.5332 [hep-ph]
44. P.P. Novichkov, J.T. Penedo, S.T. Petcov, A.V. Titov, JHEP **1907**, 165 (2019). arXiv:1905.11970 [hep-ph]
45. T. Kobayashi, Y. Shimizu, K. Takagi, M. Tanimoto, T.H. Tatsuishi, Phys. Rev. D **100**(11), 115045 (2019). [erratum: Phys. Rev. D **101**(3), 039904 (2020)]. arXiv:1909.05139 [hep-ph]
46. T. Kobayashi, Y. Shimizu, K. Takagi, M. Tanimoto, T.H. Tatsuishi, H. Uchida, Phys. Rev. D **101**(5), 055046 (2020). arXiv:1910.11553 [hep-ph]
47. R.C. Gunning, *Lectures on Modular Forms* (Princeton University Press, Princeton, NJ, 1962)
48. B. Schoeneberg, *Elliptic Modular Functions* (Springer, 1974)
49. N. Koblitz, *Introduction to Elliptic Curves and Modular Forms* (Springer, 1984)
50. J.H. Bruinier, G.V.D. Geer, G. Harder, D. Zagier, *The 1-2-3 of Modular Forms* (Springer, 2008)
51. S. Ferrara, D. Lust, A.D. Shapere, S. Theisen, Phys. Lett. B **225**, 363 (1989)
52. M. Chen, S. Ramos-Sánchez, M. Ratz, Phys. Lett. B **801**, 135153 (2020). arXiv:1909.06910 [hep-ph]
53. S. Antusch, V. Maurer, JHEP **1311**, 115 (2013). arXiv:1306.6879 [hep-ph]
54. F. Björkeroth, F.J. de Anda, I. de Medeiros Varzielas, S.F. King, JHEP **1506**, 141 (2015). arXiv:1503.03306 [hep-ph]
55. P.A. Zyla et al., [Particle Data Group], PTEP **2020**(8), 083C01 (2020)
56. M. Holthausen, M. Lindner, M.A. Schmidt, JHEP **1304**, 122 (2013). arXiv:1211.6953 [hep-ph]
57. H. Okada, M. Tanimoto, JHEP **03**, 010 (2021). arXiv:2012.01688 [hep-ph]
58. C.Y. Yao, J.N. Lu, G.J. Ding, JHEP **05**, 102 (2021). arXiv:2012.13390 [hep-ph]
59. I. Esteban, M.C. Gonzalez-Garcia, M. Maltoni, T. Schwetz, A. Zhou, JHEP **09**, 178 (2020). arXiv:2007.14792 [hep-ph]
60. S. Vagnozzi, E. Giusarma, O. Mena, K. Freese, M. Gerbino, S. Ho, M. Lattanzi, Phys. Rev. D **96**(12), 123503 (2017). arXiv:1701.08172 [astro-ph.CO]
61. N. Aghanim et al., Planck. Astron. Astrophys. **641**, A6 (2020). arXiv:1807.06209 [astro-ph.CO]
62. T. Kobayashi, H. Otsuka, Phys. Rev. D **102**(2), 026004 (2020). arXiv:2004.04518 [hep-th]
63. H. Abe, T. Kobayashi, S. Uemura, J. Yamamoto, Phys. Rev. D **102**(4), 045005 (2020). arXiv:2003.03512 [hep-th]
64. K. Ishiguro, T. Kobayashi, H. Otsuka, JHEP **03**, 161 (2021). arXiv:2011.09154 [hep-ph]

Chapter 11
Status and Overview of Neutrino Physics with Neutrino Telescopes

Luigi Antonio Fusco

Abstract The current generation of large-volume Cherenkov neutrino telescopes has shown its performance in various studies in the field of neutrino physics, from the GeV to the PeV energy scale. Both the IceCube and ANTARES Collaborations have tested neutrino oscillation with atmospheric neutrinos, consolidating our knowledge in this field. Neutrinos also offer a portal to Beyond Standard Model Physics, with some scenarios that can already be tested with the current detectors. The next-generation neutrino telescopes, IceCube-Gen2 and KM3NeT, will push forward our knowledge and understanding of the fundamental properties of neutrinos both with the atmospheric and cosmic neutrino fluxes. This contribution provides an overview of the current status of the field and the possible future results coming from neutrino telescopes.

11.1 Large-volume Cherenkov Neutrino Detectors

Neutrino telescopes are large-volume 3D arrays of photodetectors placed at large depths in a transparent medium (water or ice). Cherenkov light induced by relativistic charged particles passing through the medium is detected in the array and the recorded information can be used to reconstruct the direction and energy of the incoming neutrino, by which those charged particles were produced. Neutrino telescopes can detect all-flavour neutrino interactions [1]. Charged Current (CC) muon neutrino interactions in the medium produce a long-lived muon that can pass through, or close-by, the instrumented volume and it can be classified as a *track*: the long lever arm of the track event signature allows a precise reconstruction of the incoming neutrino direction, but the fact that only part of the track can be observed limits the precision

---

Luigi Antonio Fusco on behalf of the ANTARES and KM3NeT Collaborations.

L. A. Fusco (✉)
Dipartimento di Fisica E.R. Caianiello, Università di Salerno e INFN Gruppo Collegato di Salerno, Via Giovanni Paolo II 132, Fisciano 84084, Italy
e-mail: lfusco@unisa.it

G. Ricciardi et al. (eds.), *8th Workshop on Theory, Phenomenology and Experiments in Flavour Physics*, Springer Proceedings in Physics 292,
https://doi.org/10.1007/978-3-031-30459-0_11

of the energy reconstruction. Neutral Current (NC) interactions of neutrinos of all flavours, CC electron neutrino interactions and, for energies below some hundreds TeV, CC interactions of tau neutrinos, produce electromagnetic and hadronic particle cascades in close proximity to the neutrino interaction vertex, giving rise to the *shower* event topology: for these events an almost calorimetric measurement of the energy is possible, but the accuracy of the event direction reconstruction is limited. Finally, high-energy CC tau neutrino interactions can also produce a peculiar event topology if the tau lepton lives long enough to travel some tens of meters in the medium: 2 separate particle cascades are indeed produced, the first at the interaction vertex and the second at the decay point of the tau, giving rise to the *double-bang* event topology.

The IceCube Collaboration [2] has build the largest and most sensitive neutrino telescope on Earth, at a depth between 1.4 and 2.4 km below the ice surface in the Antarctic ice sheet—approximately at the geographic South Pole. The IceCube km^3-scale detector has been operated continuously in its full configuration since 2010. The first observation of high-energy (TeV-to-PeV) cosmic neutrinos came out of its data [3] and has been later confirmed in other analyses with different datasets [4–6]. Inside IceCube, the DeepCore infill—which uses the same technology as the main IceCube detector—is maily used to study atmospheric neutrinos below 100 GeV. The ANTARES Collaboration [7] deployed and operated for 15 years, from 2007 to 2022, a neutrino telescope in the Mediterranean Sea, at a depth between 2 and 2.45 km below sea level in front of the French coast, 40 km off-shore Toulon. Even though much smaller than IceCube, ANTARES had been able to obtain significant results in various analyses concerning both neutrino physics and astronomy, especially when searching for neutrinos below 100 TeV from the Southern Sky.

The next-generation neutrino telescope in the Mediterranean Sea, KM3NeT [8], is currently being built and has already shown its potential in the analysis of the first data. KM3NeT will allow studying neutrino astrophysics with the ARCA detector, a km^3 neutrino detector located at a maximal depth of 3.5, 100 km off-shore the Sicilian coast, and neutrino oscillation physics with the ORCA detector, which will be located off-shore the coasts of France, close to the ANTARES site, covering a total volume of $\sim$0.007 km^3. Also at the South Pole, the IceCube Collaboration is planning to improve its detector, by both adding a denser infill that will boost its capabilities in the GeV energy range—the IceCube Upgrade [9]—and enlarging its size to target PeV cosmic neutrinos with IceCube-Gen2 [10].

The physics goals of these detectors are broad: from the study of very-high-energy astrophysics to the fundamental properties of the Standard Model. The former will be described in Sect. 11.2, while the latter will be covered in Sect. 11.3 of this document. An outlook on the future of the field will be given in Sect. 11.4.

11.2 Astrophysics with Neutrino Telescopes

Neutrino astrophysics is a branch of astroparticle physics that searches for high-energy neutrinos in order to pinpoint the sources and accelerators of primary Cosmic Rays (CRs) or the properties of their propagation through the Cosmos [1]. Neutrino telescopes are optimised to search for high-energy (TeV-to-PeV) extraterrestrial neutrinos. In order to do so, they must first eliminate the foregrounds coming from neutrinos and muons produced by CRs interacting in the Earth atmosphere. Generally, CR muons can be removed by using the Earth as a shield when looking for particles coming from below the detector. The atmospheric neutrino flux becomes less relevant with increasing energies: thus, selecting high-energy neutrinos enhances the probability of picking-up any cosmic ones. This general concept is not always followed since interesting sources might be in the sky above the neutrino telescope: vetoing techniques can be used to get rid of the atmospheric downward-going foregrounds if the detector is large enough. Following this more sophisticated approach, the IceCube collaboration was able to obtain the first evidence of the presence of cosmic neutrinos by searching for High-Energy Starting Events [3]. Further analyses, using both tracks [4] and shower [5] events have provided a highly-statistically-significant confirmation of this observation. This signal is compatible with an isotropic diffuse flux of cosmic neutrinos, with only a few associations to possible source currently available [6, 11]. The ANTARES Collaboraiton has also done similar analyses, using both tracks and showers. Even though non-significant, a small excess of events of possible cosmic origin is observed in ANTARES data, above the expectation from the atmospheric foregrounds [12]. In doing so, both the IceCube and ANTARES collaboration have provided an estimation of the normalisation and spectral index of the cosmic flux, which is typically assumed to follow a simple power-law behaviour. A compilation of the results obtained by both detectors, taken from Ref. [13], is shown in Fig. 11.1. Looking at the results from the different analyses, it is possible to see that a slight tension is present between the different fit results. The reason behind this tension might be multifaceted: the different IceCube samples cover different parts of the sky, across different energy ranges and with different atmospheric foregrounds due to the fact that the atmospheric neutrino flux is flavour-dependent. At the moment it is not possible to give a conclusive answer to this tension, but several hypothesis have been brought up related to both the astrophysics of the neutrino sources, neutrino propagation effects and possible signals beyond the standard model.

Given these observations, it can be useful to study the cosmic neutrino flavour ratio, that is the ratio between electron, muon and tau neutrinos in the observed signal. The flavour ratio at Earth of high-energy neutrinos that have travelled over cosmic distances carries valuable information, both for what concerns their sources and the physics of neutrino oscillations. According to the simplest emission model, cosmic neutrinos are produced in the decay chains of pions coming from proton-proton (pp) or proton-photon ($p\gamma$) interactions. In the case of these pion-decay scenarios, the neutrino flavour ratio at their source is expected to be $\nu_e : \nu_\mu : \nu_\tau = 1 : 2 : 0$. Because of oscillations, this is transmuted on Earth into a $1 : 1 : 1$ ratio [15]. In

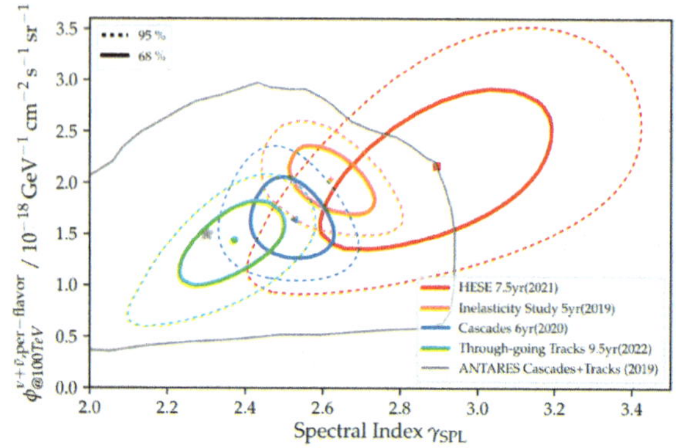

Fig. 11.1 Overview of astrophysical diffuse neutrino flux measurements to date, taken from Ref. [13]. The plot shows the 68–95% confidence level contours from the analysis of up-going tracks in IceCube with 9.5 years of data taking [4], the IceCube high-energy starting events [3], IceCube contained cascades [5], IceCube starting tracks [14] and ANTARES using both cascades and tracks with 9 years of data [12]

different scenarios, for example if neutrinos are produced from neutron decay, a pure electron neutrino beam is emitted by the source; neutrino oscillation change this into a ∼0.55 : 0.17 : 0.28 flavour ratio. Even if no tau neutrino is emitted from the source, oscillations will produce a possibly detectable amount of tau neutrinos in the detectors on Earth. Since the atmospheric tau neutrino flux is negligible in the 100 TeV–PeV energy range, even a single tau neutrino detection would be a statistically significant evidence of a cosmic signal. The IceCube Collaboration has claimed the observation of 2 candidate τ neutrino events in the cosmic diffuse sample by measuring the time profile of the measured hits in proximity of the vertex [16]. It is thus possible to constrain the neutrino flavour ratio at Earth and, assuming standard neutrino oscillations, the flavour ratio at the sources. This is shown in Fig. 11.2 where the flavour content of the IceCube signal is fitted in terms of its three components and compared to the expectations from different models at source. The uncertainty on this measurement is still large and more statistics is needed to better constrain any emission scenario.

11.3 Neutrino Physics with the Atmospheric Beam

Cosmic rays impinging on air nuclei at the top of the atmosphere generate secondary particles in the *cosmic-ray air shower*. Leptons typically emerge from semileptonic decays of mesons, such as pions and kaons, in the air shower. The resulting atmospheric neutrino flux is the most abundant neutrino signal in neutrino telescopes. The measurement of the energy spectrum of the atmospheric neutrino flux is extremely valuable in every neutrino telescope, since it provides a direct way to study those

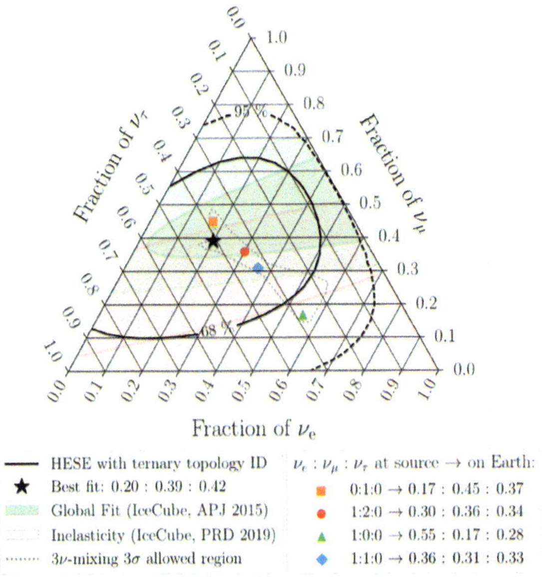

Fig. 11.2 Neutrino flavour composition of the IceCube HESE events in terms of the ν_e, ν_μ and ν_τ content in the cosmic signal. Contours show the 1 and 2σ confidence intervals. Flavor compositions expected from various astrophysical neutrino production mechanisms are marked, and the entire accessible range of flavor compositions assuming standard 3-flavor mixing is shown. Figure taken from Ref. [16]

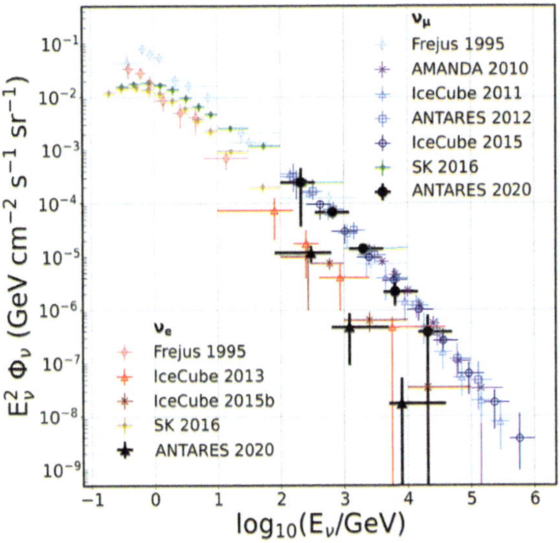

Fig. 11.3 Measured energy spectra of the atmospheric ν_e and ν_μ flux in neutrino observatories, as reported in the legend. Error bands show the statistical plus systematical uncertainties in the flux measurement. The figure is taken from Ref. [18], where the full list of references of all measurements can be found

systematic effects coming from the detector which mainly affect the energy estimation. This has been done in several analyses in ANTARES [17, 18] and IceCube [19, 20], with an overview of all measurements obtained in the GeV–PeV range given in Fig. 11.3. In the plot, both electron and muon neutrino spectra are shown, with all instruments giving similar outcomes.

At low energy, below 100 GeV, the atmospheric neutrino flux flavour composition is affected by the phenomenon of neutrino oscillations [15]. Typically, muon

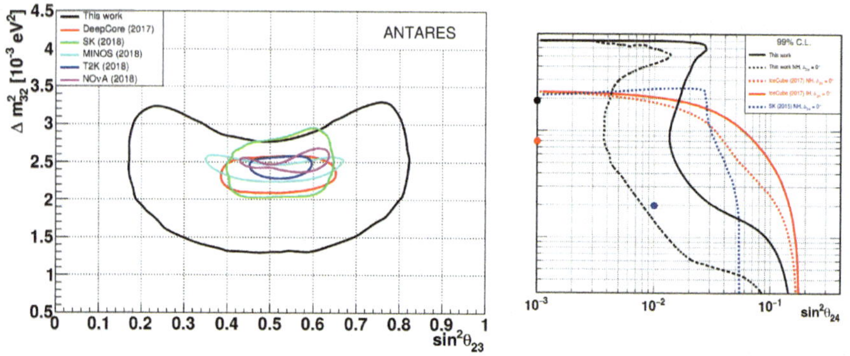

Fig. 11.4 Left: 90% confidence level contours in the plane $sin^2\theta_{23}$–Δm^2_{23} obtained by different neutrino experiments. Right: 99% confidence level contours in the search for neutrino oscillations to a fourth, sterile, neutrino family, in the plane $sin^2\theta_{24}$–$sin^2\theta_{34}cos^2 2\theta_{24}$. More details on the various samples are provided in Ref. [23], from which this figure has been extracted

neutrino disappearance can be measured in neutrino telescopes, i.e. the lack of muon neutrino events in the upward-going track with respect to the expectations from the models describing the neutrino production in the atmosphere. These muon neutrinos would mainly transform into tau neutrinos driven by the θ_{23} and Δm^2_{23} parameters of the neutrino mixing matrix. Neutrino telescopes, even though optimised for neutrino astronomy, can still access this low-energy realm since the atmospheric flux is very abundant there, and large statistics of events can be collected. To boost its performance, the IceCube Collaboration has put under the ice a more densely-instrumented infill of the detector, DeepCore [21]. This lowered the neutrino detection energy threshold to a few GeV allowing accessing the energy region most affected by oscillations [22]. Even though also the ANTARES detector was not optimised for the task, it still managed to collect enough neutrino events to measure the neutrino oscillation parameters and also constrain the possible existence of a sterile neutrino into which standard model neutrinos could oscillate [23]. The results obtained by IceCube and ANTARES on this subject are shown in Fig. 11.4.

11.4 Outlook for the Next Generation Neutrino Telescopes

The discoveries by IceCube have shown that neutrino telescopes can play a significant role in the field of both astroparticle and particle physics. ANTARES has also shown that using sea-water as a detection medium is viable. When comparing water and ice, different phenomena affect light propagation. The deep Antarctic ice is in general more transparent than sea water, while photon scattering is less evident in water with respect to ice. The former property allows the detection of a larger quantity of photons for equivalent events; however most of these photons will be deflected when arriving at the photosensors in ice, thus losing part of the timing and directional information they carried at their emission point. As a consequence, the neutrino properties can be reconstructed much better in water than in ice. For this reason, an under-water neutrino telescope is the optimal instrument to further push the researches carried out in the field. The KM3NeT Collaboration has started the construction of the two instruments that will possibly provide answers to the unknowns in both neutrino astrophysics and neutrino oscillation physics. The outlook for these topics will be described in the next pages of this document.

11.4.1 Cosmic Searches

Even though the IceCube cosmic diffuse signal is clearly evident, not a single source of cosmic neutrinos (and thus of cosmic rays) as been identified with a 5σ significance. The flaring blazar TXS 0506+056 has been detected with a 3σ significance by the IceCube collaboration in coincidence with gamma-ray observations from MAGIC and Fermi-LAT [6], and similarly the star-forming Galaxy NGC1068 is a hot-spot in the IceCube neutrino sky [11], but the significance of this observation is just at about 4σ, too [24]. A better angular resolution is necessary to pinpoint the sources of neutrinos: indeed, assuming that detector efficiencies are the same, the sensitivity of a neutrino telescope scales linearly with the improvement of the angular accuracy in the reconstruction of the neutrino incoming direction. While IceCube can reach a few tenth-of-degree precision for some high-energy track events, KM3NeT/ARCA aims at reconstructing 100 TeV neutrinos with at 0.1° uncertainty, thus boosting significantly the reach of the instrument [25]: using water as the detection medium is the optimal choice for neutrino astronomy. To do so, in KM3NeT/ARCA more than $1\,\mathrm{km}^3$ of Mediterranean sea water will be instrumented with 230 Detection Units (DUs), flexible strings anchored to the sea-bed holding 18 Digital Optical Modules (DOMs). In each DOM, 31 3-inch photomultiplier tubes are installed, to precisely measure the Cherenkov light. Figure 11.5 shows the expected sensitivity in the search for cosmic neutrino sources as a function of the source declination of the ARCA detector compared with the current world's best results from IceCube and ANTARES. Also shown are the results coming from the analysis of about 3 months of the partial ARCA detector, with 6 active DUs [26].

11.4.2 Neutrino Oscillations with Atmospheric Neutrinos

The oscillations of atmospheric neutrinos crossing the Earth are influenced by the presence of electrons in the matter [27, 28]. Depending on the Neutrino Mass Ordering (NMO), neutrinos and antineutrinos are affected differently. Cherenkov neutrino detectors cannot distinguish between neutrinos and antineutrinos, but the atmospheric flux and the neutrino-matter interaction cross sections are not the same for neutrinos

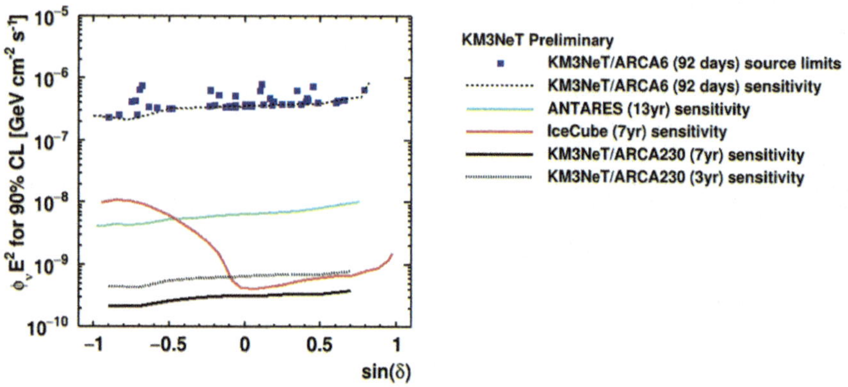

Fig. 11.5 Upper limits on point-like sources of neutrinos with a differential spectral index proportional to E^2 from the analysis of the first data taken with 6DU configuration of ARCA for 93 days (blue dots) [26] compared to the sensitivity obtained by ANTARES and IceCube and the expectations for the full KM3NeT ARCA detector [25], all shown with lines according to the legend

and antineutrinos: as a consequence, if a large-enough amount of events of all flavours is collected, the NMO could be measured by observing the oscillation patterns in energy and zenith observed in the detector [29]. The KM3NeT ORCA detector will follow this strategy with the goal of determining the NMO with a significance of 3–5σ, to be reached in a few years of data taking—depending on the actual NMO [30]—as shown in Fig. 11.6 (left). The detector is currently being build and taking data; the final goal is to instrument ~6.7 Mton of sea water with 115 DU. Each DU will be roughly 200m-long and will be placed on the seabed with an inter-DU horizontal spacing of ~20 m. The first analysis with a very partial detector configuration (only 6 DUs and 1 year of data taking) has already provided evidence of muon neutrino disappearance because of oscillations, with a first estimation of the atmospheric neutrino oscillation parameters that is already compatible with the results obtained by other experiments [31], as reported in Fig. 11.6 (right).

Given its very large volume, compared to all other atmospheric neutrino experiments, KM3NeT/ORCA will collect a very large number of neutrino interactions. In particular, being sensitive to tau neutrino appeararance (namely as an excess of events for the shower event topology), the normalisation of the tau component could be measured precisely, testing the unitarity of the neutrino mixing [30]. Three years of data taking are found to be sufficient to exclude deviation from unitarity at a level of 20% at 3σ level.

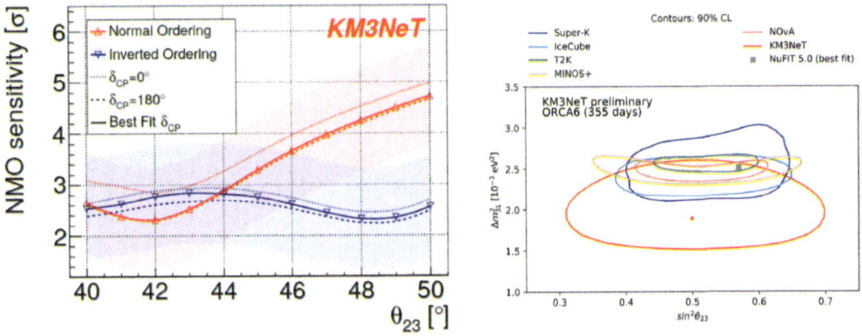

Fig. 11.6 Left: expected significance in the determination of the NMO with 3 years of data taking with the complete ORCA detector as a function of the true value of the atmospheric mixing angle, assuming normal or inverted ordering, for different values of the amplitude of the charge-parity violation phase [30]. Right: contour at 90% confidence level with the first data of ORCA for the atmospheric oscillation parameters. Contours of other experiments have been added for comparison purposes as well as the global oscillation-fit best fit value [31]

11.5 Conclusions

Neutrino telescopes, originally built with the intention to discover cosmic ray accelerators have also shown that they can give valuable inputs in various fields of particle and astroparticle physics. ANTARES and IceCube have been operated now for 15 years and a greet wealth of results came from them in the search for astrophysical neutrinos, neutrino flavour physics and also (not reported in this document) beyond standard model physics. The next generation of neutrino telescopes in the Mediterranean Sea, KM3NeT/ARCA and KM3NeT/ORCA, will further push forward the reach of these instruments aiming at determining the origin of cosmic rays and the neutrino mass ordering. First encouraging results have already been produced with the first data from KM3NeT and the future looks promising for the physics output that will come out of it.

References

1. T. Gaisser, R. Engel, E. Resconi, *Cosmic Rays and Particle Physics*, 2nd edn. (Cambridge University Press, Cambridge, 2016)
2. https://icecube.wisc.edu/
3. R. Abbasi et al., PRD **104**, 022002 (2021)
4. R. Abbasi et al., ApJ **928**, 50 (2022)
5. M.G. Aartsen et al., PRL **125**, 121104 (2020)
6. M. Aartsen et al., Science **361**(6398), 147–151 (2018)
7. https://antares.in2p3.fr/
8. S. Adrián-Martnez et al., J. Phys. G **43**(8), 084001 (2016)
9. A. Ishihara, PoS (ICRC2019)1031 (2019)

10. M.G. Aartsen et al., J. Phys. G **48**(6), 060501 (2021)
11. R. Abbasi et al., Science **378**(6619), 538–543 (2022)
12. L.A. Fusco, F. Versari, POS(ICRC2019)891 (2019)
13. M. Ackermann et al., (2022). arXiv:2203.08096 [hep-ph]
14. M.G. Aartsen et al., PRD **99**, 032004 (2019)
15. M. Spurio, Particles and Astrophysics (Springer, 2015)
16. R. Abbasi et al., arXiv:2011.03561 [hep-ex], 2020
17. S. Adrián-Martinez et al., EPJC **73**, 2606 (2013)
18. A. Albert et al., PLB **816**, 136228 (2021)
19. R. Abbasi et al., PRD **83**, 012001 (2011)
20. M.G. Aartsen et al., EPJC **75**, 116 (2015)
21. R. Abbasi et al., Astropart. Phys. **35**(10), 615–624 (2012)
22. M.G. Aartsen et al., PRL **120**, 071801 (2018)
23. A. Albert et al., JHEP **2019**, 113 (2019)
24. M.G. Aartsen et al., PRL **124**, 051103 (2020)
25. S. Aiello et al., Astropart. Phys. **111**, 100–110 (2019)
26. R. Muller et al., (2022). https://doi.org/10.5281/zenodo.6805393
27. S.P Mikheyev et al., Sov. J. Nucl. Phys. **42**(6), 913–917 (1985)
28. L. Wolfenstein, PRD **17**(9), 2369–2374 (1978)
29. E.K. Akhmedov et al., JHEP **02**, 082 (2013)
30. S. Aiello et al., EPJC **82**, 26 (2022)
31. L. Nauta et al., POS(ICRC2021)1123 (2021)

Chapter 12
Neutrino Oscillations in T2K and Prospects of the Hyper-Kamiokande Experiment

Joanna Zalipska and Hyper-Kamiokande

Abstract Nowadays T2K is providing us with precise measurement of oscillation parameters and search for CP violation in neutrino physics. This work describes recent analysis of the T2K data corresponding to 1.97×10^{21} POT of υ beam and 1.63×10^{21} POT of υ beam after introducing new selections and updating flux prediction and neutrino interaction models. The results suggests the upper octant of θ_{23}, while lower octant is still allowed at 68% C.L. We have found that CP conserving values lays outside of 90% C.L., when the best fit shows δ_{CP} around $-\pi/2$. We also obtained mild preference for normal mass ordering. This extensive work will be continued by Hyper-Kamiokande experiment, which physics programme and status are discussed.

12.1 Introduction

12.1.1 Where We Are in Neutrino Physics Measurements?

Measurement performed in neutrino physics provided very interesting results in last years. During last two decades not only existence of neutrino oscillations was proved [1, 2], but also all oscillation parameters have been measured. Therefore we already know values of all mixing angles and mass square differences [3].

At this moment we are concentrating on the precise measurements of oscillation parameters. Current experiments are trying to answer the question weather θ_{23} mixing angle is maximal—we would like to learn if $\sin^2 \theta_{23}$ is greater than 0.5. On the other hand, from solar neutrino measurements we know than $m_2{}^2$ is larger than $m_1{}^2$, but we are still trying to answer the question what is mass ordering for atmospheric sector, is $\Delta m^2{}_{23} > 0$? However, the most interesting question for all neutrino physicists is what is value of δ_{CP}, *0*, π or in between? Current accelerator experiments such as

J. Zalipska (✉) · Hyper-Kamiokande
National Centre for Nuclear Research, A. Sołtana 7, 05–400 Otwock, Poland
e-mail: joanna.zalipska@ncbj.gov.pl

G. Ricciardi et al. (eds.), *8th Workshop on Theory, Phenomenology and Experiments in Flavour Physics*, Springer Proceedings in Physics 292,
https://doi.org/10.1007/978-3-031-30459-0_12

T2K [4, 5] or NovA [6] are trying to answer those questions and new experiments such as Hyper-Kamiokande [7] and DUNE [8] are planned to be built in near future to perform new interesting measurements in neutrino physics.

12.1.2 What We Measure in the T2K Experiment?

The T2K accelerator experiment is using muon neutrino or antineutrino beam to study neutrino or anti-neutrino oscillations. It can probe muon neutrino disappearance, $\nu_\mu \rightarrow \nu_\mu$, as well as electron neutrino appearance, $\nu_\mu \rightarrow \nu_e$. Simplified equations which describes those oscillations are presented here:

$$P(\nu_\mu \rightarrow \nu_\mu) \approx 1 - 4cos^2\theta_{13}sin^2\theta_{23} \cdot \left(1 - cos^2\theta_{13}sin^2\theta_{23}\right) \cdot sin^2\left(1.27\Delta m^2_{32}L/E\right)$$

$$P(\nu_\mu \rightarrow \nu_e) \approx sin^2(2\theta_{13})sin^2\theta_{23} \cdot sin^2\left(1.27\Delta m^2_{32}L/E\right) \mp sin^2\left(1.27\Delta m^2_{21}L/E\right) \cdot 8J_{CP} \cdot sin^2\left(1.27\Delta m^2_{32}L/E\right)$$

where Jarlskog invariant is defined as following:

$$J_{CP} = 1/8cos\theta_{13}sin(2\theta_{12})sin(2\theta_{23})sin(2\theta_{13}) \cdot sin\delta_{CP} \approx 0.033 \cdot sin\delta_{CP}$$

Those equations do not include matter effect. They rely on mixing angles θ_{13}, θ_{23}, θ_{12} and mass square differences $\Delta m^2{}_{32} = m^2{}_3 - m^2{}_2$, $\Delta m^2{}_{21} = m^2{}_2 - m^2{}_1$, as well as unknown δ_{CP} violating phase. Oscillation probability depends also on the quantities which are characteristic for designed experiment such as: energy of produced neutrinos—*E* and distance which neutrinos travel between production and detection point—*L*.

The energy spectrum of muon and electron neutrinos for the T2K experiment are shown on Fig. 12.1 together with oscillation probabilities. When we concentrate on $\nu_\mu \rightarrow \nu_\mu$ oscillations the dip of the oscillation is driven by the value of the $sin^2 2\theta_{23}$ while position of oscillation maximum depends on $\Delta m^2{}_{32}$. Similarly when we consider $\nu_\mu \rightarrow \nu_e$ oscillation the dip of oscillation depends on the value of θ_{13} and its sensitive to octant of θ_{23}. However, this effect is combined with effect introduced by the value of the δ_{CP} violating phase—which is different for neutrinos and anti-neutrinos, and matter effect.

The T2K experiment is measuring $\nu_\mu \rightarrow \nu_e$ and $\nu_\mu \rightarrow \nu_\mu$ oscillations using samples in the far detector characteristic for interactions of muon and electron neutrinos performing simultaneous fit of those samples.

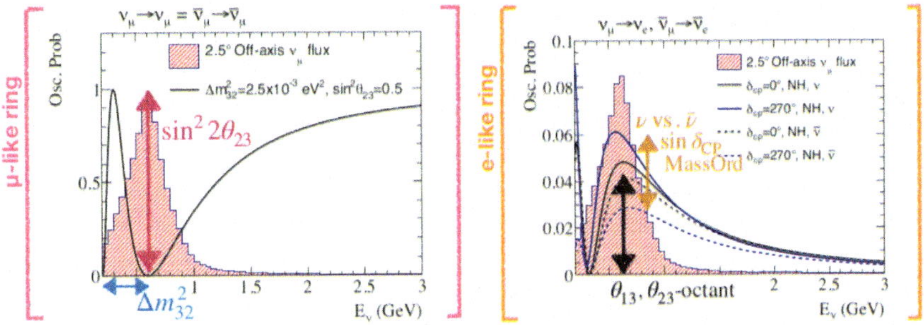

Fig. 12.1 Oscillated spectrum of muon and electron neutrinos on left and right plot respectively. The oscillation probability curve is also shown

12.2 The T2K Accelerator Experiment

T2K is a long baseline accelerator neutrino experiment located in Japan which started its operation in 2010 [2]. It produces neutrino beam with peak energy around 600 meV using accelerator in the JPARC laboratory. The accelerator produces 30 GeV proton beam which is sent on graphite target producing hadrons such as π and K. Those particles are focused by horns. Further they decay producing neutrino beam. When the horn current is reversed then hadrons of opposite sign are focused and in consequence the anti-muon neutrino beam is created.

280 m from target the near detectors are placed. We have two near detectors: INGRID—scintillator detector placed on the axis of the neutrino beam and the Near Detector (ND280) which is located at 2.5 degrees off-axis at the direction to the far detector. ND280 is composed of the scintillator Figne Grained Detectors interleaved with Time Projection Chambers and placed inside a magnet. Those detectors are used to monitor beam direction and constrain neutrino beam flux as well provide us with measurements of the cross-sections of neutrino interactions.

Then neutrino beam is travelling 295 km where at 2.5 degrees off-axis the far detector-Super-Kamiokande (SK) is placed [9]. The SK detector is a tank located 1 km underground in Japanese Alps which is filled with 50 kT of ultra pure water. Inside the tank the structure is placed which holds photodetectors. Here we use about 11,146 20″ photomultiplier tubes (PMT) facing inside the detector and about 1885 8″ PMTs which are looking outside the inner volume. The detector is measuring Cherenkov radiation produced by charged particles created in interactions of neutrinos. Light creates ring patterns on the wall of the detector which are measured by PMTs. By studying shape of the ring we can distinguish muons from electrons and therefore we can tell if the interacting neutrino was υ_μ or υ_e. Thanks to this fact we can study oscillations in two modes. On the other hand by measuring time of detected signal in each PMT we can reconstruct direction of the interacting neutrino.

12.3 Oscillation Analysis

12.3.1 Analysis Overview

To measure oscillations we perform the likelihood analysis which compares the observed data in the Super-Kamiokande far detector and predictions based on the models. The T2K performs two simultaneous fits of the υ_e and υ_μ events using frequentist and Bayesian approaches. Both analyses start with neutrino flux prediction which relis on beam line simulation and hadron production model. Then the near detector data are used to constrain flux and interaction uncertainties, which as input used modes of neutrino interactions. To estimate the oscillation parameters the far detector fit is performed. The near detector fit is done separately or together with the far detector fit depending on the analysis.

12.3.2 What's New in the 2022 Oscillation Analysis?

This year we introduced few following changes in the analysis:

- updated flux prediction,
- updated neutrino interaction model,
- introduced new proton and photon tagging in the near detector,
- introduced new μ-like CC1π sample in the far detector.

The hadron production on the graphite target is measured by NA61/SHINE experiment at CERN [10]. Recently the new hadron production data taken at 2010 were analysed. They contain more statistics for π^+ production and adds K^+ and proton production data and additional updates on the part of the models including cooling water in horns. In consequence the new systematic error on the produced flux was assigned as can be seen in Fig. 12.2. If one compares the black dashed line from Fig.12.2 representing previous results with continuous black line representing current analysis then they can see that in most energy regions the systematic error on neutrino flux for υ_μ and υ_e modes were reduced.

The interaction model update includes improved treatment for Spectral Function model and additional uncertainties for resonant (RES) and multi-pion (CC multi-π) events as well as Final State Interactions.

In the analysis of the near detector data ND280 the new selections of proton tracks and photons were introduced. It helps us to split previously used sample of CC0π events into two samples depending on the number of reconstructed proton tracks: the CC0π-0p-0γ and CC0π-Np-0 γ. This increases ability to constrain charged current quasi-elastic (CCQE) and 2p2h (represented by Meson Exchange Current) interactions when performing fit of the ND280 data. Selection of photons affected mostly samples of CC1π and CC-Other used for previous analysis which are dominated by RES and deep inelastic interactions (DIS). Using this new introduced selection

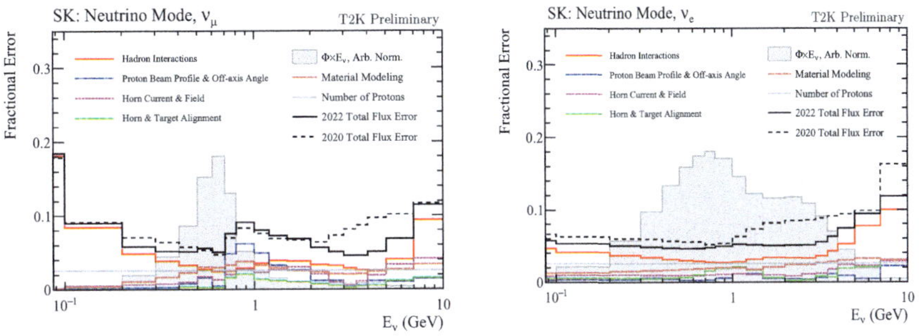

Fig. 12.2 Flux prediction uncertainty for Super-Kamiokande site for υ_μ and υ_e at left and right plot respectively

relying on photon tagging allowed us to split those two samples into CC1π-0γ, CCPhoton and CCOther-0 γ, which created new samples dominated by DIS and CCπ^0.

By performing the fit to near detector data the uncertainty of the prediction of flux in the far detector was substantially reduced as can be seen from Fig. 12.3—compare uncertainty before the fit marked in red with flux and cross-section uncertainty after the fit marked in blue.

In the far detector Super-Kamiokande the new multi-ring sample was introduced. This sample tags CC1π^+ events from υ_μ interactions. New sample contains events with one reconstructed muon ring and two decay electrons and events with two muon rings and one decay electron. Decay electron signal can be produced by electron from muon decay which was produced directly in neutrino interaction or which is a product of charged pion decay. It is marked as a delay signal after detecting main muon ring. This increases muon-like sample of neutrino interactions by about 30%. This selected sample is sensitive to oscillations, but in higher energies than nominal sample. Its due to the fact that this sample is dominated by CC multi-π and CC-DIS events.

12.3.3 Data Samples Used for Analysis

In current analysis we use the same data set as it was used for analysis in 2020 [11]. It uses data from Run 1 to Run 10 taken from 2010 to 2020. The latest data of Run 11 taken in 2021 has not been used in the analysis so far. Those data were taken after adding Gadolinium to the water in Super-Kamiokande detector. We are still working on the analysis of that data set.

Summary of the data statistics used by Near Detector and Far Detector for neutrino and anti-neutrino modes are summarized in Table 12.1.

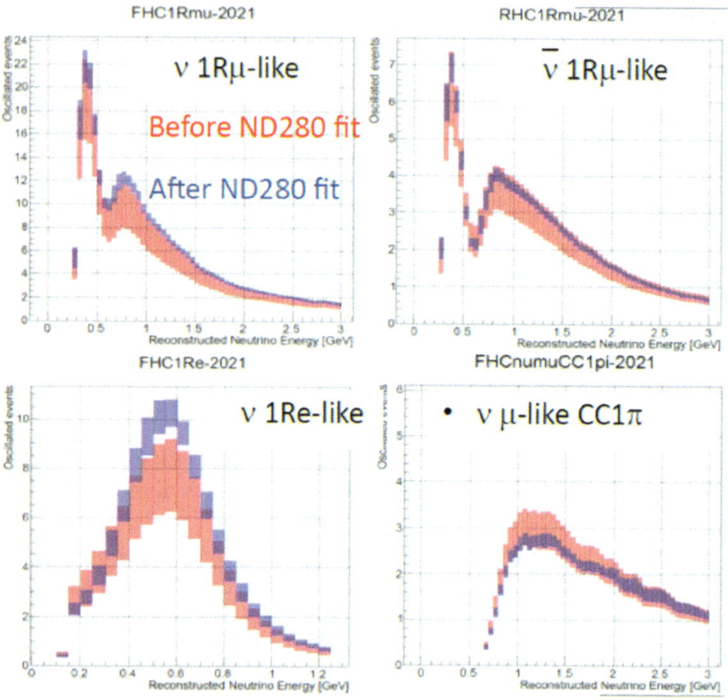

Fig. 12.3 *T2K Preliminary*: Predictions of oscillated spectra of neutrinos at Super-Kamiokande with errors before and after ND280 fit marked in red and blue respectively. The upper row shows μ-like events for neutrino at left and anti-neutrino beam at right. The bottom left plot shows *e*-like events from neutrino beam while the bottom right plot shows new sample of μ-like events CC1π

Table 12.1 Data set used for current analysis

Sample	Near detectors	Far detector
υ mode	1.39×10^{21} POT	1.97×10^{21} POT
$\bar{\nu}$ mode	0.63×10^{21} POT	1.63×10^{21} POT

12.4 Oscillation Analysis Results

12.4.1 Detected Samples

Events in the far detector are classified depending on the number of rings and type of the reconstructed ring *e*-like or μ-like. As it was mentioned before, so far we were using single ring samples for the analysis and this year we added multi-ring sample of $CC1\pi^+$ events. Summary of numbers of detected events is presented in Table 12.2 together with number of events expected in each sample under assumption of different values of δ_{CP} violating phase. For the event numbers listed in the table the other oscillation parameters were set to the values from previous analysis from

Table 12.2 Summary of data and MC events. MC was generated setting oscillation parameters from results of 2020 analysis and assuming different values of δ_{CP}

Sample	$\delta = -\pi/2$	$\delta = 0$	$\delta = \pi/2$	$\delta = \pi$	Data
υ, 1Re	102.7	86.7	71.1	87.1	94
υ, 1Re CC1π	10.0	8.7	7.1	8.4	14
υ, 1Rμ	379.1	378.3	379.1	380.0	318
υ, MRμ CC1π^+	116.5	116.0	116.5	117.0	134
υ, 1Re	17.3	19.7	21.8	19.4	16
υ, 1Rμ	144.9	144.5	144.9	145.3	137

2020. It is worth to comment that the 1Ring *e*-like CCπ^+ sample, which was already used for analysis in 2020, contains events with invisible charged pion based on the detection of decay electron.

From Table 12.2 we can see that detected number of events for υ_e samples are consistent with Monte Carlo prediction for $\delta_{CP} = -\pi/2$. One can also see that new sample of υ_μ CC1π interactions increases statistics of υ_μ samples of about 30%.

12.5 Results for Atmospheric Sector

The joint fit of $\upsilon_\mu \rightarrow \upsilon_\mu$ and $\upsilon_\mu \rightarrow \upsilon_e$ for neutrino and anti-neutrino runs was performed simultaneously fitting Near and Far Detectors data using Bayesian method (as well as frequentists method). It allows us to derive oscillation parameters as shown in Fig. 12.4 for atmospheric region. The Normal and Inverted Orderings are shown separately. For the fit of the T2K data with setting reactor constraint on the θ_{13} parameter the best fit of θ_{23} is in the upper octant. However, the lower octant is still allowed in 68% C.L.

12.5.1 Results of δ_{CP} and θ_{13}

As a result of the same fit the result of electron neutrino appearance were obtained. The two dimensional contours of δ_{CP} versus $\sin^2\theta_{13}$ are shown on Fig. 12.5 separately for Normal and Inverted mass Orderings. The 68% C.L. regions based on the T2K fit only (upper plots) show that the CP conserving values are excluded at 68% C.L. The contour becomes more restrictive when the constraint on θ_{13} from reactor experiments is used in the fit (bottom plots).

Figure 12.6 shows comparison of the posterior probability of the Bayesian analysis, which results have been discussed so far, with the other oscillation analysis using frequentists approach. Both analyses show that the CP conserving values lay outside of the 90% C.L. and the best fit is close to maximal CP violation of $-\pi/2$.

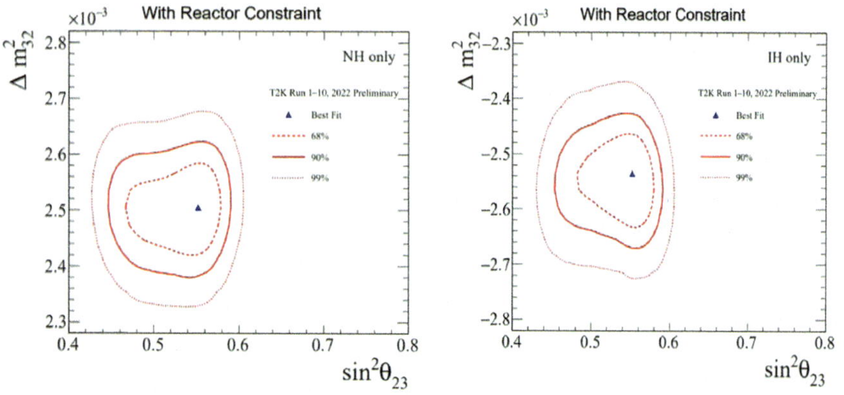

Fig. 12.4 Allowed regions of atmospheric parameters for Normal Ordering and Inverted Ordering shown in left and right plot respectively

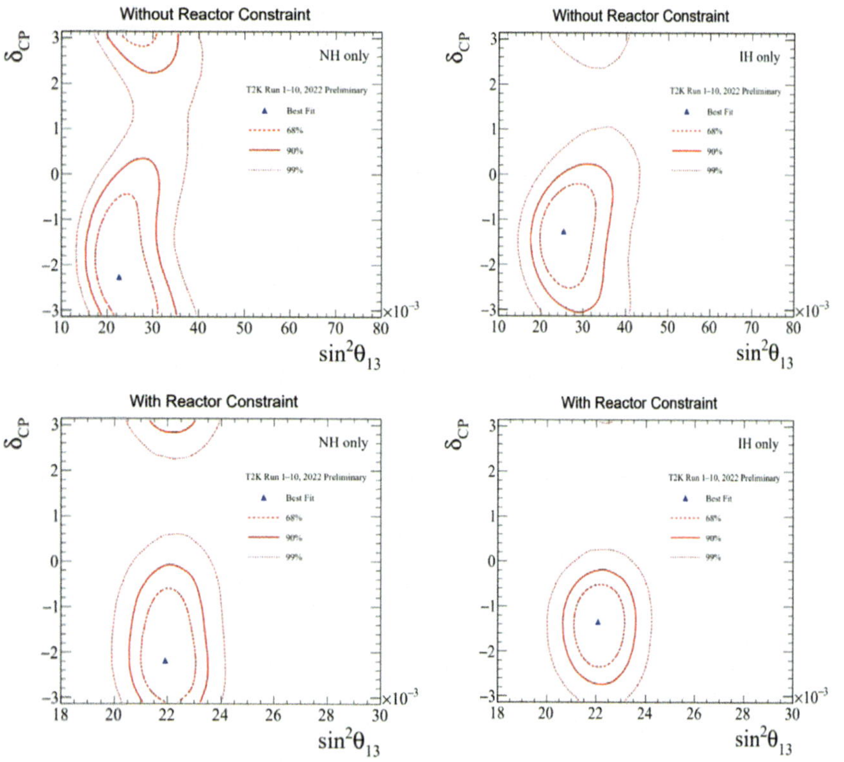

Fig. 12.5 Allowed regions of δ_{CP} versus $\sin^2\theta_{13}$ for Normal and Inverted mass Orderings on left and right plots respectively. The upper row shows results of the fit without reactor constraint on θ_{13} value, while the bottom plots shows results with reactor constraint

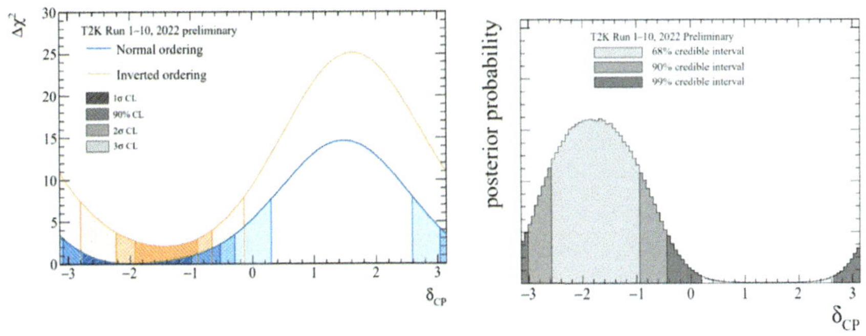

Fig. 12.6 Left plot: $\Delta\chi^2$ versus δ_{CP} obtained for frequentists analysis, the blue line shows Normal Ordering while the red line shows Inverted Ordering. Right plot: Posterior probability versus δ_{CP} for Bayesian analysis, here results are marginalized over mass orderings

Table 12.3 Posterior probabilities for Bayesian analysis for different combinations of octant and mass ordering hypotheses

T2K-only	T2K preliminary	$\sin^2\theta_{23} < 0.5$	$\sin^2\theta_{23} > 0.5$	Sum
	NO ($\Delta m^2_{32} > 0$)	0.24	0.39	0.63
	IO ($\Delta m^2_{32} < 0$)	0.15	0.22	0.37
	Sum	0.39	0.61	1.00
T2K + reactor	T2K preliminary	$\sin^2\theta_{23} < 0.5$	$\sin^2\theta_{23} > 0.5$	Sum
	NO ($\Delta m^2_{32} > 0$)	0.20	0.54	0.74
	IO ($\Delta m^2_{32} < 0$)	0.05	0.21	0.26
	Sum	0.25	0.75	1.00

We present also posterior probabilities for the different combinations of octant and mass ordering hypotheses in Table 12.3. It is necessary to emphasize that those results have limited significance. The sum listed in rows of 0.61 for *T2K-only* fit and sum of 0.75 for *T2K + reactor* suggests preference of the upper octant, $\sin^2\theta_{23} > 0.5$. The sum of posterior probabilities listed in columns of 0.63 and 0.74 for *T2K-only* and *T2K + reactor* respectively, shows preference of Normal Ordering.

12.5.2 Future Prospects of Oscillation Analysis

There are two analyses which T2K experiment is currently working on.

The first one is the joint fit of the T2K beam and the Super-Kamiokande atmospheric neutrino data. Super-Kamiokande atmospheric covers wide range of the T2K energies, but it has in particular sensitivity for mass ordering for higher energies. The sensitivity study of this analysis has been performed giving the ability to exclude the CP conservation as function of δ_{CP} shown in Fig. 12.7. We can gain some sensitivity

Fig. 12.7 Sensitivity to measure δ_{CP} for SK atmospheric analysis only in black, T2K analysis only in blue and joint T2K + SK atmospheric analysis in red

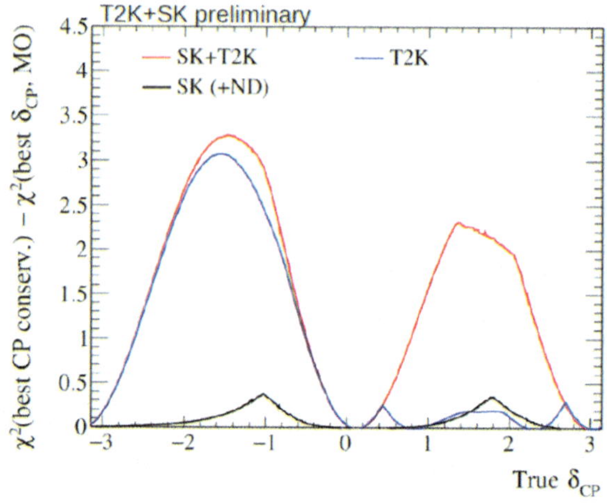

for values of δ_{CP} below *0* comparing red (joint fit) with blue (T2K only) curves. On the other hand, we can gain a lot of sensitivity for values of δ_{CP} greater that zero.

We are also working on combined effort of two collaborations to perform joint fit of the T2K and Nova experiments data. In this case situation is more difficult since we are trying to combine data from two different long baseline experiments which use different baselines (T2K–295 km, Nova–810 km) and different neutrino energies (peak energy for T2K—0.6 GeV, for Nova—2 GeV). On top of it both experiments use different detector technologies, T2K—water Cherenkov detector, Nova–segmented liquid scintillator bars. Currently, the collaborations are working on sensitivity study and breaking degeneracy between mass ordering and δ_{CP}.

12.6 Future of the T2K Experiment

Since experiment is working since 2010, then some upgrades related to the operation of the T2K are ongoing. We are working to upgrade main ring magnets and horn current supplies. It will allow operation at higher current 250 to 320 kA and provide higher intensity beam. It is expected to be ready in 2023.

We are also working on the upgrade of the Near Detector 280. One component of the detector, called PI0 detector is going to be replaced by scintillator detector SuperFGD laying horizontally, and placed between two horizontal Time Projection Chambers (HTPC). HTPCs have almost the same design as currently used TPCs. The SuperFGD is newly designed detector which contains 2 million 1 cm^3 scintillator cubes read out by optical fibres. This highly segmented detector will allow to lower threshold for proton and charged pion detection. The horizontal layout will improve efficiency for reconstruction of high angle tracks. In general those improvements will help to measure cross-sections of neutrino interactions. The SuperFGD detector is already in Japan and its expected to operate in 2023.

The last upgrade of the T2K experiment is related to the Super-Kamiokande detector. SK is loaded in Gadolinium sulfate since summer 2020 [12] and recently we added more Gadolinium to the water. Gadolinium helps with neutron tagging. Neuton is captured by Gd which emits gammas, and therefore allows for discrimination between neutrino and anti-neutrino interactions. T2K already recorded data in the SK-Gd phase, but they are not yet analized. We are still working on the analysis, but we already can see neutron capture signal in the data.

12.7 The Hyper-Kamiokande Experiment

Future of neutrino physics in Japan is related to planned Hyper-Kamiokande experiment (HK) [7]. The Hyper-Kamiokande collaboration is a group of more than 400 physicists from 20 different countries. HK is going to be scaled up version of the Super-Kamiokande water Cherenkov detector. History of the water Cherenkov detectors in Japan started in 80s of last century when the Kamiokande detector of 3 kT volume was built. After successful measurements performed by the Kamiokande experiment the Super-Kamiokande started its operation in 1996. As it was mentioned before the SK tank contains 50 kT of water. Now, in order to perform more precise measurements we are planning to build Hyper-Kamiokande detector which volume will be 260 kT, so 5 times larger than current SK detector mass.

12.7.1 Detectors of Hyper-Kamiokande

The HK detector will be also located in Japanese Alps but in different mine than currently running SK [13]. This location is also compatible with 2.5 degrees off-axis beam sent from JPARC accelerator centre in Tokai. The baseline of the experiment will be also 295 km, as for T2K. The detector tank will have diameter of 68 m and high of 71 m Fig. 12.8, and it will be filled with 260 kT of ultra pure water. It will have 20% photocathod coverage assured by 20$''$ PMTs with better performance than current one used for SK together with newly designed multi-PMT units. Current status of work shows that tunnel excavation is almost done and the cavern excavation will start soon. The tank itself will be built in 2024/2025.

The Hyper-Kamiokande collaboration is also planning to build Intermediate Water Cherenkov Detector Fig.12.8 [14]. This detector is planned to be located 1 km from the target. The aim of the detector is to measure unoscillated neutrino flux for different off-axis angles in order to reduce uncertainties related to predicted spectrum of neutrinos in the far detector. It will be 1 kT scale water Cherenkov detector doped with gadolinium with minimum overburden. For this purpose the water tank of 8 m diameter and 6 m high will be built. It will move inside a chimney of 50 m height

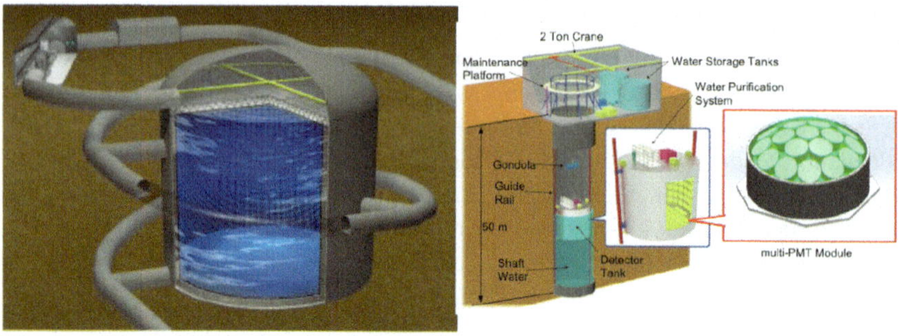

Fig. 12.8 Left: Scheme of the Hyper-Kamiokande detector. Right: Scheme of the Intermediate Water Cherenkov Detector

to measure neutrinos for different off-axis angles, see Fig. 12.8. It will be equipped with multi-PMT units.

About 20,000 new 20" PMTs will be built for the HK detector. The production of those PMTs has already started and will be finished by 2026. Those new inner detector PMTs will have single photon detection efficiency of 24%, so twice more than for current PMTs used by SK. The dark noise rate will be at the level of 4 kHz, while the time resolution will be 1.5 ns comparing to 3 ns for PMTs from SK currently in use. Those better performance will improve reconstruction of the neutrino interaction event.

On top of it new multi-PMT units are designed. Multi-PMT is an array of 19 3" PMTs as can be seen on the right scheme in Fig. 12.8. They are planned to be used in HK together with new 20" PMTs and in Intermediate Water Cherenkov Detector. The dark noise will be 19 $\times$ 200–300 Hz and transit time spread of 1.3 ns. Pixelizing of the multi-PMT will provide directional information and improve spatial and timing resolution.

12.7.2 *Physics Programme of Hyper-Kamiokande*

The HK has wide physics programme [15]. However the main goal is precise measurement of neutrino oscillations and CP violation phase, as well as determination of mass ordering. It will be probed using atmospheric neutrinos and neutrino/anti-neutrino beam from JPARC.

For this purpose the sensitivity study was performed assuming 10 years of data taking, which corresponds to 2.10×10^{22} *POT*, for 1:3 neutrino to anti-neutrino beams composition. Figure 12.9 show discovery potential for true value of δ_{CP} for Normal Mass Ordering. As one can see a wide range of δ_{CP} can be covered (blue line), with some improvements when atmospheric data are included in the analysis (black line). Generally atmospheric neutrinos can help to remove mass ordering ambiguity.

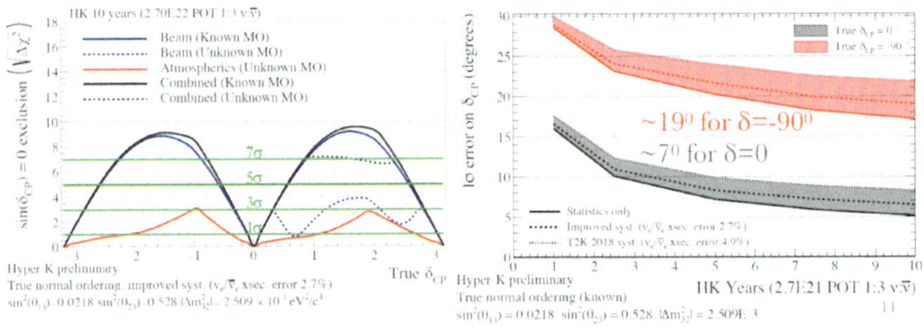

Fig. 12.9 Left: Sensitivity plot for δ_{CP} for 10 years of Hyper-Kamiokande operation. Right: Precision of δ_{CP} measurement as a function of the year of operation of HK

Precision of measurement of two values of δ_{CP} 0° and 90° is presented in Fig. 12.9 as function of the year of HK data taking. One can see that after 10 years of operation we can reach 7° precision if δ_{CP} equals 0 and 19° precision of δ_{CP} equals 90°. Therefore HK will have good chance to discover CP violation.

Except neutrino oscillations the HK will search for proton decay. It's worth to mention that it was the main purpose to build Kamiokande detector in 80s, but as you know since then we extended limits on proton lifetime. The huge water tank of HK containing a lot of protons will allow to extend current limits by one order of magnitude for such channels as $p \to e^+ + \pi^0$, $p \to n + K^+$.

On the other hand, the HK will continue to look for supernovae burst. We will also continue to study solar neutrinos and search for dark matter.

To summarize, currently we are at the level of preparation of the HK experiment, some production of HK components as PMTs has already started. We are planning to start operation of the HK in 2027.

12.8 Conclusions

The T2K experiment is currently performing precise measurements of oscillation parameters, such as θ_{23} and $\Delta m^2{}_{32}$ and looking for CP violation and determination of mass ordering. This programme will be continued by the Hyper-Kamiokande experiment. Except oscillation physics its programme includes search for nucleon decay, solar neutrino measurements, astrophysics neutrinos and dark matter search.

Recently the T2K analysis was enhanced with new samples allowing to exclude CP symmetry conservation at 90% C.L. and obtain mild preference for upper octant and normal mass ordering.

In future we are planning to use T2K-Gd data, which has already been collected as well as perform joint analysis of the T2K + SK-atmospheric and T2K + Nova data. Ongoing are upgrades of the neutrino beam and Near Detector.

The HK will take over the T2K job from 2027 when we hope to see a lot of interesting measurements.

References

1. The Super-Kamiokande Collaboration, Phys. Rev. Lett. **81**, 1562–1567 (1998)
2. The SNO Collaboration, Phys. Rev. Lett. **89**(1), 011301 (2002)
3. R.L. Workman et al. (Particle Data Group), Prog. Theor. Exp. Phys. **2022**, 083C01 (2022)
4. T2K Homepage, http://t2k.org, https://t2k-experiment.org/
5. The T2K Collaboration, *Nucl.Instrum.Meth.A* 659 (2011) 106–135
6. Nova Homepage, https://novaexperiment.fnal.gov/
7. Hyper-Kamiokande Homepage, https://www.hyperk.org/
8. DUNE Homepage, https://www.dunescience.org/
9. The Super-Kamiokande Collaboration, Nucl. Instrum. Meth. A **501**, 418–462 (2003)
10. NA61/SHINE Homepage, https://shine.web.cern.ch/
11. The T2K Collaboration, Phys. Rev. D **103**(11), 112008 (2021)
12. The Super-Kamiokande Collaboration, Nuclear Inst. Method. Phys. Res. A 1027, 166248 (2022)
13. K. Abe et al., February 2016, KEK preprint https://lib-extopc.kek.jp/preprints/PDF/2016/1627/1627021.pdf
14. M. Scott, arXiv:1603.01251
15. K. Abe et al., The Hyper-Kamiokande Collaboration, arXiv:1502.05199

Chapter 13
Constraining Extended Scalar Sectors at Current and Future Colliders—An Update

Tania Robens

Abstract In this proceeding, I discuss several models that extend the scalar sector of the Standard Model by additional matter states. I here focus on results for models with singlet extensions, which have been obtained recently and update some of the results presented in previous work. In more detail, I will briefly review the option to test a strong first-order electroweak phase transition using precision measurements in the electroweak sector, as well as production cross-sections for non-standard scalar production at Higgs factories.

13.1 Introduction

In this proceeding, I discuss various new physics models that extend the Standard Model (SM) by adding additional fields that transform as singlets under the SM gauge group. The proceedings contain results which can be seen as follow-ups of the studies presented e.g. in [1]. I will therefore focus on novel developments within the last year. In particular, I discuss

- The real singlet extension of the SM, which comes with an additional scalar that transforms as a singlet. The model features one additional CP even neutral scalar. See [2–6] for original literature as well as [7, 8] for results presented here;
- The two real singlet extension (TRSM), where the SM scalar sector is extended by two additional gauge singlets, featuring in total three CP even neutral scalars that also allow for interesting cascade decays. Original literature can be found in [9], while the results discussed here have first been presented in [10].

All models are confronted with most recent theoretical and experimental constraints. Theory constraints include the minimization of the vacuum as well as the

---

RBI-ThPhys-2022-36.

T. Robens (✉)
Rudjer Boskovic Institute, Bijenicka cesta 54, 10000 Zagreb, Croatia
e-mail: trobens@irb.hr

G. Ricciardi et al. (eds.), *8th Workshop on Theory, Phenomenology and Experiments in Flavour Physics*, Springer Proceedings in Physics 292,
https://doi.org/10.1007/978-3-031-30459-0_13

requirement of vacuum stability and positivity. We also apply constraints from perturbative unitarity and perturbativity of the couplings at the electroweak scale.

Experimental bounds include the agreement with current measurements of the properties of the 125 GeV resonance discovered by the LHC experiments, as well as agreement with the null-results from searches for additional particles at current or past colliders. We additionally impose constraints from electroweak precision observables (via S, T, U parameters [11–13]). For our studies, we use a combination of private and public tools. In particular, we use `HiggsBounds` [14–17] and `HiggsSignals` [18, 19] for the comparison with current collider constraints. Experimental numbers are taken from [20, 21] for electroweak precision observables. Some predictions for production cross sections shown here have been obtained using Madgraph5 [22].

13.2 Real Singlet Extension

As a first simple example, we discuss a real singlet extension of the SM with a $\mathbb{Z}_2$ symmetry previously reported on in [2–6, 23]. The $\mathbb{Z}_2$ symmetry is softly broken by a vacuum expectation value (vev) of the singlet field, inducing mixing between the gauge-eigenstates which introduces a mixing angle α. The model has in total 5 free parameters. Two of these are fixed by the measurement of the 125 GeV resonance mass and electroweak precision observables. The free parameters of the model are then given by

$$\sin\alpha,\ m_2,\ \tan\beta\ \equiv\ \frac{v}{v_s}, \tag{13.1}$$

with v (v_s) denoting the doublet and singlet vevs, respectively. We concentrate on the case where $m_2\ \geq\ 125\,\text{GeV}$, where SM decoupling corresponds to $\sin\alpha\ \rightarrow\ 0$.

13.2.1 *Current Constraints*

Limits on this model are shown in Fig. 13.1, including a comparison of the currently maximal available rate of $H\ \rightarrow\ h_{125}h_{125}$ with the combination limits from ATLAS [24] as well as dedicated searches with full run II data [25–27].[1] The most constraining direct search bounds are in general dominated by searches for diboson final states [29–32]. In some regions, the Run 1 Higgs combination [33] is also important. Especially [31, 32] currently correspond to the best probes of the models parameter space.[2]

---

[1] I thank D. Azevedo for providing exclusion limits for $b\bar{b}\tau^+\tau^-$ [27] in a digitized format, as used in [28].

[2] We include searches currently available via `HiggsBounds`.

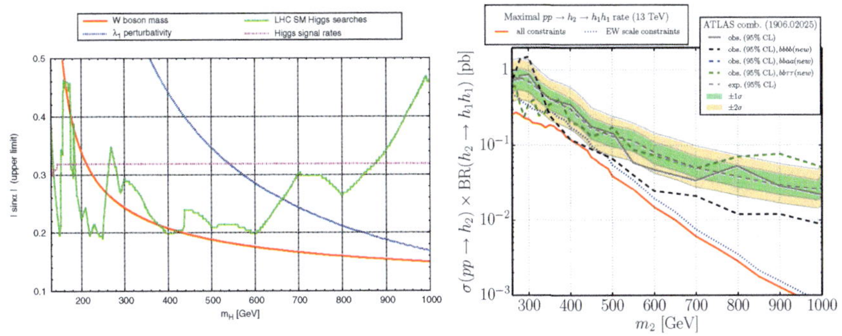

Fig. 13.1 Results for the singlet extension, *Left:* comparison of current constraints for a fixed value of $\tan\beta = 0.1$, taken from [7]. *Right:* maximal $H \to h\,h$ allowed, with electroweak constraints at the electroweak scale (blue) or including RGE running to a higher scale (red), in comparison with results from the ATLAS combination and direct searches. Taken from [8]

13.2.2 Strong First-Order Electroweak Phase Transition

One question currently of high interest is whether new physics will enable a strong first-order electroweak phase transition. In general, this is studied in the context of exotic Higgs decays, see e.g. [34] for recent work within the Snowmass process. However, one can also test this in models where the second scalar is heavier than the 125 GeV resonance, as e.g. discussed in [35, 36].

In [37], we have investigated whether regions in the models parameter space that allow for a strong first-order electroweak phase transition can be tested using electroweak precision measurements, in particular, the Higgs signal strength as well W-boson mass, where we resort to the PDG value [38] $m_W^{\text{exp}} = 80.379 \pm 0.012\,\text{GeV}$.[3] In our study the $\mathbb{Z}_2$ symmetry is not imposed, leading to additional terms in the potential with respect to the singlet scenario discussed above.

In general, constraints depend not only on the mass and mixing angle, but also additional couplings in the potential that govern scalar self-interactions. However, two constraints that only depend on the second scalar mass and the mixing angle[4] are the one-loop corrections to the W-boson mass, as well as the signal strength measurements for the 125 GeV scalar. The latter is related to the mixing angle via $\cos^2\theta \gtrsim \mu$. In [37], we compared to the by that time current ATLAS combination value [40], $\mu = 1.06 \pm 0.06$, leading to $|\sin\theta| \lesssim 0.24$ at 95% C.L.[5] We display this bound as "current bound" in Fig. 13.2. Furthermore, the precision that might be

[3] This corresponds to the PDG value at the time of the above reference. The current value [39] is slightly lower. We do not expect this to have a qualitatively large impact.

[4] At the order of perturbation theory discussed here; extending to higher orders might introduce additional parameter dependencies.

[5] Note that the Run 2 combinations of ATLAS [41] and CMS [42] separately lead to $|\sin\theta| \lesssim 0.26$ and $|\sin\theta| \lesssim 0.33$, respectively.

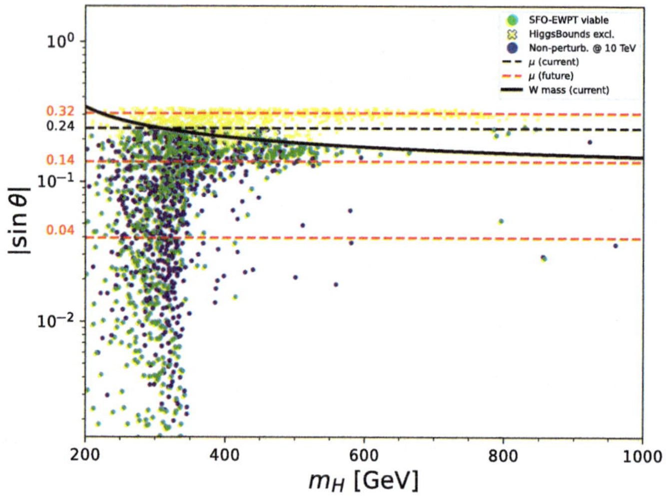

Fig. 13.2 The viable points for a SFO-EWPT are shown in lime green filled circles. The points denoted by yellow crosses yield the necessary conditions for a SFO-EWPT, but are excluded by direct searches for heavy scalars (imposed via the `HiggsBounds` package). The points denoted by blue circles are those allowed by direct searches for heavy scalars but that become non-perturbative at 10 TeV. The current W-boson mass constraint is shown in solid black and the constraints due to signal strength measurements of the SM-like Higgs are shown in dashed lines, with black indicating the current constraint at 95% C.L. ($|\sin\theta| = 0.24$) and red indicating the corresponding future constraints assuming 5, 1, and 0.1% measurements with a central value of $\mu = 1$ ($|\sin\theta| = 0.32, 0.14, 0.04$, respectively). Figure taken from [37]

achievable at future colliders for the signal strength varies from collider to collider and can reach per-mille level at future machines [43, 44]. This is shown in lines corresponding to 95% C.L., for an assumed precision of 5, 1 and 0.1% in Fig. 13.2. For these, set $\mu = 1$. For direct searches, we only impose *current* constraints through the `HiggsBounds` package.

For the calculation of the contributions to the W-boson mass, we essentially follow the work presented in [45]. However, since then many of the parameters used in the evaluation of the SM-like contribution have been updated, we have re-evaluated the SM-prediction [46] using the most recent electro-weak parameters [38, 47], leading to the theoretical prediction [37] $m_W^{\text{SM}} = 80.356$ GeV.

We see that even for permill-precision of the coupling strength, viable parameter points in this model remain which can lead to a strong first-order electroweak phase transition.

13.3 Two Real Singlet Extension

I now turn to a model where the scalar sector of the SM has been augmented by two real scalar fields, which obey a $\mathbb{Z}_2 \otimes \mathbb{Z}_2'$ symmetry. This model has first been presented in [9], with recent reviews in [10, 48].

While previous studies were concentrating on physics at hadron colliders, I now want to report on the options to investigate the model at future Higgs factories, along the lines of models discussed in [49]. In particular, I enhance the benchmark planes presented in the above references to a general scan and show production cross sections achievable at an e^+e^- collider with a center-of-mass (com) energy of 250 GeV.[6]

The particle content of this model in the scalar sector consists of 3 CP-even neutral scalars, where mass eigenstates correspond to admixtures of the gauge eigenstates. In the following, we will use the following mass hierarchy

$$M_1 \leq M_2 \leq M_3$$

and denote the corresponding physical mass eigenstates by h_i. Gauge and mass eigenstates are related via a mixing matrix. The model contains in total 9 free parameters, out of which 2 are fixed by the observation of a scalar particle with the mass of 125 GeV as well as electroweak precision observables. Apart from the masses, also the vaccum expectation values and mixing angles serve as input parameters. We scan the parameter space using the implementation into `ScannerS` [50, 51].

13.3.1 Production of Light Scalars at Higgs Factories

In this model, the only feasible production is Zh radiation of the lighter scalar, with production cross sections given in Fig. 13.3. Cross sections have been derived using Madgraph5 [22].

We can now investigate what would be production cross sections for scalar particles with masses $\lesssim$ 160 GeV at Higgs factories. Rates for the production of a 125 GeV scalar with subsequent decays into novel scalar states have been presented in [10, 48] and will not be repeated here. Instead, we turn to the Higgs-Strahlung production of new physics scalars, where we first present results for the benchmark planes presented in [9]. If we require production rates of $Z h_i$ to be larger than $\sim$ 10 fb, only BPs 4 and 5 render sufficiently large rates for the production of h_2 and h_3, respectively. Production rates are independent of the other scalars, and we therefore depict them for both BPs in Fig. 13.4.

BP4 is constructed in such a way that as soon as the corresponding parameter space opens up, the $h_1 h_1$ decay becomes dominant; final states are therefore mainly

---

[6] The parameter scans include only current bounds, not possible discovery or exclusion at e.g. a HL-LHC.

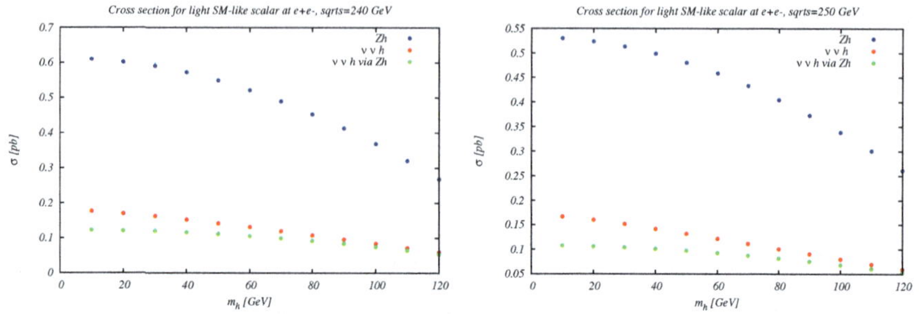

Fig. 13.3 Leading order production cross sections for $Z\,h$ and $h\,\nu_\ell\,\bar{\nu}_\ell$ production at an $e^+\,e^-$ collider with a com energy of 240 GeV *(left)* and 250 GeV *(right)* using Madgraph5 for an SM-like scalar h. Shown is also the contribution of $Z\,h$ to $\nu_\ell\,\bar{\nu}_\ell\,h$ using a factorized approach for the Z decay. Taken from [49]

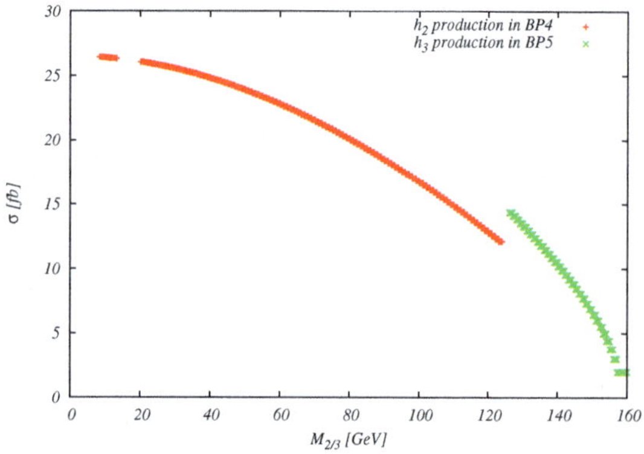

Fig. 13.4 Production cross sections for $Zh_{2/3}$ in BPs 4 and 5, respectively, at a 250 GeV Higgs factory. Taken from [10]

$Z\,b\bar{b}b\bar{b}$ if $M_2 \gtrsim 2\,M_1$. Below that threshold, dominant decays are into a $b\,\bar{b}$ pair, which means that standard searches as e.g. presented in [52, 53] should be able to cover the parameter space.

Similarly, in BP5 the $h_3 \rightarrow h_1\,h_1$ decay is also favoured as soon as it is kinematically allowed. Therefore, in this parameter space again $Zb\bar{b}b\bar{b}$ final states become dominant. Otherwise $Z\,b\bar{b}$ and $Z W^+ W^-$ final states prevail, with a cross over for the respective final states at around $M_3 \sim 135\,\text{GeV}$. Branching ratios for these final states are in the 40–50% regime.

13.3.2 More General Scan

In the following, I present results for a more general parameter scan of the masses, taken from [49, 54], given in Fig. 13.5. In this figure, two data-sets are considered which fulfill all current constraints as implemented using the current versions of

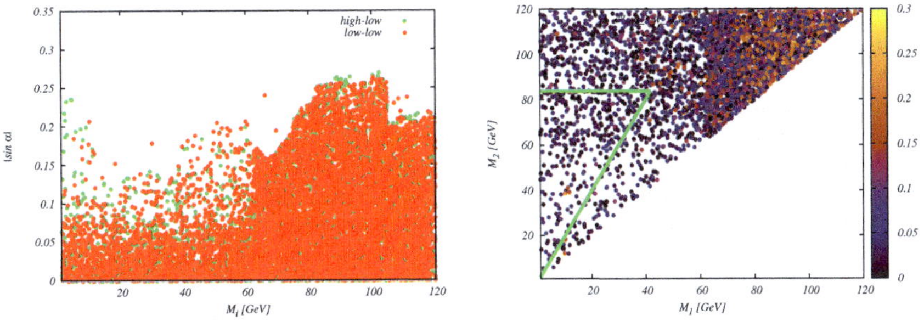

Fig. 13.5 Available parameter space in the TRSM, with one (high-low) or two (low-low) masses lighter than 125 GeV. *Left*: light scalar mass and mixing angle, with $\sin\alpha = 0$ corresponding to complete decoupling. *Right:* available parameter space in the $(M_1,\ M_2)$ plane, with color coding denoting the rescaling parameter $\sin\alpha$ for the lighter scalar h_1. Within the green triangle, $h_{125} \rightarrow h_2 h_1 \rightarrow h_1\, h_1\, h_1$ decays are kinematically allowed. Taken from [49]

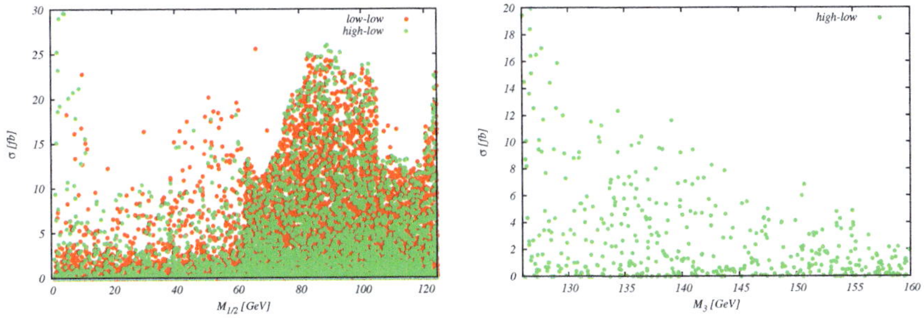

Fig. 13.6 Maximal production cross section for Higgs-Strahlung for scalars of masses $\neq$ 125 GeV in the TRSM for points passing all discussed constraints. Production cross sections depend on the parameter point and can reach up to 30 fb. Taken from [49]

`ScannerS` and `HiggsBounds`, `HiggsSignals`. The are labelled "low-low" if both $M_{1,2} \leq 125\,\text{GeV}$ and "high-low" if $M_1 \leq 125\,\text{GeV},\ M_3 \geq 125\,\text{GeV}$.

Here, $|\sin\alpha|$ is symbolic for the respective mixing angle, with $\sin\alpha = 0$ denoting complete decoupling. We see that in general, for low mass scalars, mixing angles up to ∼0.3 are still allowed. At Higgs factories, it was shown in [49] that $Z\,h$ production is dominant in the low mass range, and also gives the largest contribution to the $\nu\bar{\nu}h$ final state, so we concentrate on Higgs-strahlung. Maximally allowed production cross sections at an e^+e^- collider with a com energy of 250 GeV are displayed in Fig. 13.6.

As branching ratios for the low mass scalars are inherited via mixing with the scalar from the SM-like doublet, the largest production cross sections are obtained for scenarios where the light scalars decay into $b\bar{b}$ final states; these could in principle be investigated or constrained using already existing projecting bounds at Higgs factories, see discussion in [49]. We display cross sections for such final states in

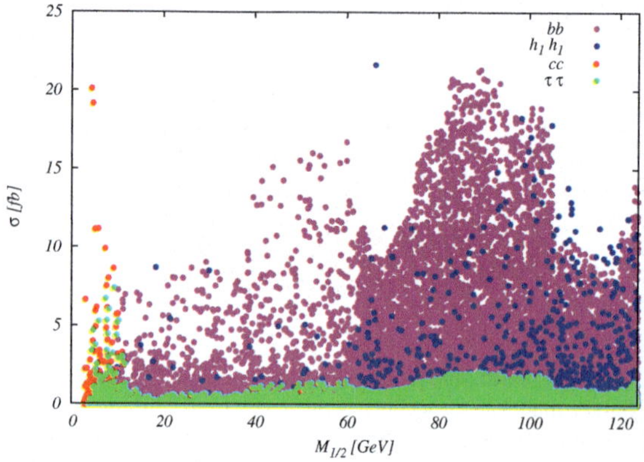

Fig. 13.7 Production cross sections for $e^+e^- \to Z h_{1/2} \to Z X X$, with $X \equiv b$ *(magenta)* and h_1 *(blue)*. Points from all data sets are included. Cross sections can reach up to 20 fb. In the low mass region, also $X \equiv \tau, c$ final states can become important. Taken from [49]

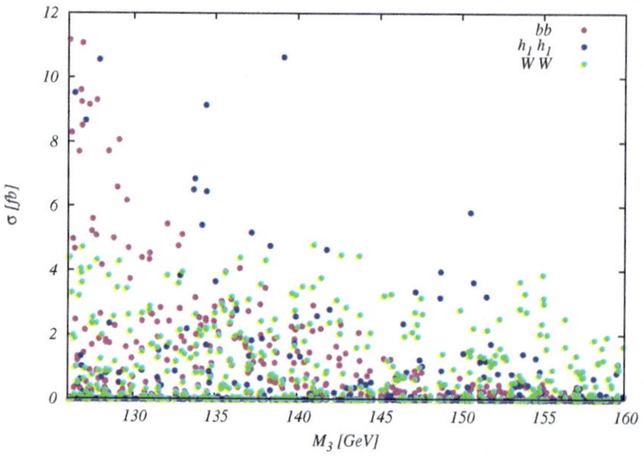

Fig. 13.8 Production cross sections for $e^+e^- \to Z h_3 \to Z X X$, with $X \equiv b$ *(magenta)*, h_1 *(blue)*, and W *(green)*. Points from all data sets are included. Cross sections can reach up to ~12 fb. Taken from [49]

Fig. 13.7, together with predictions for $h_1 h_1$ final states in case $h_2 \to h_1 h_1$. For $M_i \lesssim 12$ GeV, other final states as e.g. $\tau\tau$ and $c\bar{c}$ can lead to cross sections up to 20 fb.

For scenarios with $M_3 \gtrsim 126$ GeV, $h_1 h_1$, $W^+ W^-$, and $b\bar{b}$ can dominate, depending on the specific parameter point. We display the corresponding production cross sections in Fig. 13.8.

Finally, I display cross sections $e^+e^- \to Z h_{2/3}$, with subsequent decays to $h_1 h_1$ final states in Fig. 13.9. We find the largest cross section of about ~ 20 fb for a parameter point where $M_2 \sim 66$ GeV, $M_1 \sim 18$ GeV. The h_1 in this parameter point decays predominantly into $b\bar{b}$ final states with a branching ratio of about 85%.

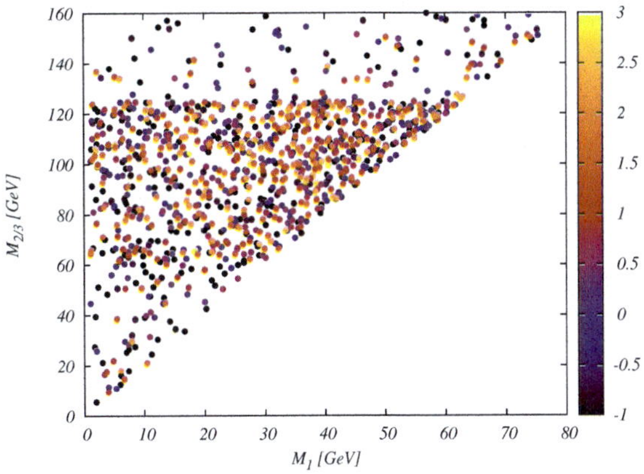

Fig. 13.9 Production cross sections for $e^+e^- \rightarrow Z\,h_{2/3} \rightarrow Z\,X\,X$, with $X \equiv h_1$, in the $(M_1,\ M_{2/3})$ plane. Color coding refers to the $\log_{10}[\sigma/\text{fb}]$ for Zh_1h_1 production. Maximal cross sections are around 20 fb. Taken from [49]

13.3.3 Experimental Searches with TRSM Interpretations

I also want to comment on experimental searches at the LHC that have made use of TRSM interpretations. Both were performed by the CMS experiment and consider asymmetric production $H_3 \rightarrow h_1\,h_2$, with subsequent decay into $b\bar{b}b\bar{b}$ final states [55], as well as $b\bar{b}\gamma\gamma$ in [56]. For this, maximal production cross sections were provided in the parameter space, allowing all additional new physics parameter to float; the respective values have been tabulated in [57, 58]. In Figs. 13.10 and 13.11, the expected and observed limits in these searches are displayed for the TRSM and NMSSM [59].

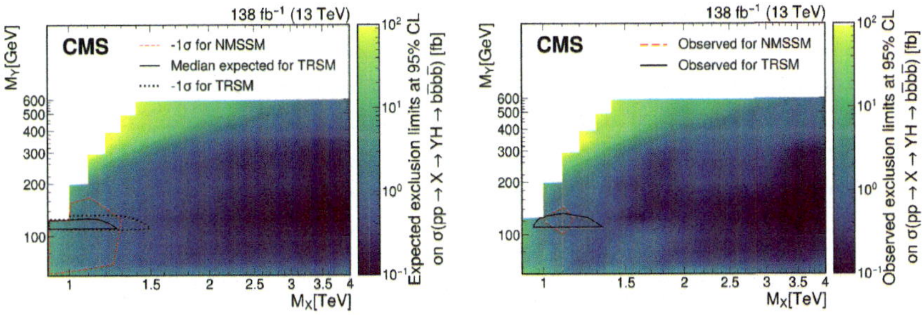

Fig. 13.10 Expected *(left)* and observed *(right)* 95% confidence limits for the $p\,p \rightarrow h_3 \rightarrow h_2\,h_1$ search, with subsequent decays into $b\bar{b}b\bar{b}$. For both models, maximal mass regions up to $m_3 \sim 1.4\,\text{TeV}$, $m_2 \sim 140\,\text{GeV}$ can be excluded. Figure taken from [55]

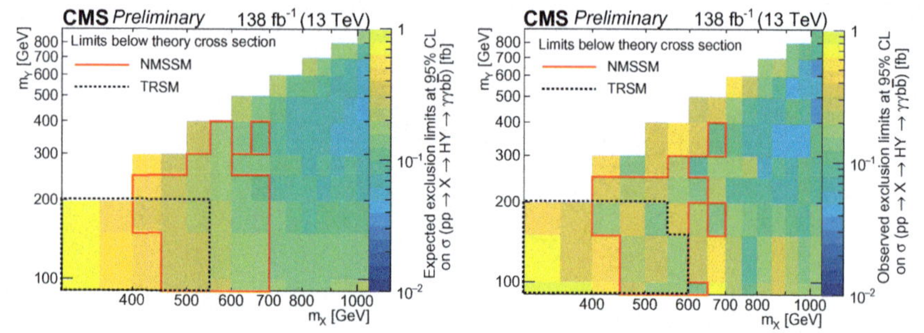

Fig. 13.11 Expected *(left)* and observed *(right)* 95% confidence limits for the $p\,p \rightarrow h_3 \rightarrow h_2\,h_1$ search, with subsequent decays into $b\bar{b}\gamma\gamma$. Depending on the model, maximal mass regions up to $m_3 \sim$ 800GeV, $m_2 \sim$ 400 GeV can be excluded. Figure taken from [56]

Several additional searches also investigate decay chains that can in principle be realized within the TRSM, as e.g. other searches for the same final states [60] or $b\,\bar{b}\mu^+\mu^-$ [61] final states.

13.4 Summary and Outlook

In these proceedings, I discussed novel results which build on the models presented in [1], which were in the same form presented at the Capri Workshop. In particular, I focussed on scenarios with singlet extensions in the SM scalar sector. For a simple real singlet extension, I showed updated results in the $(M_H;\,|\sin\alpha|)$ plane as well as a comparison of possible maximal rates with current bounds for di-Higgs production. I also commented on the possibility to constrain regions in the models parameter space that allow for a strong first-order electroweak phase transition via future precision measurements in the electroweak sector.

The second model I presented is the TRSM, where the scalar sector is augmented by two additional scalars that obey a $\mathbb{Z}_2 \times \mathbb{Z}_2'$ symmetry. Here, I showed predictions for light scalar production at future Higgs factories, as well as some sample results of current LHC searches interpreted within this model.

In summary, the discussion and investigation of BSM scenarios is ongoing. The models discussed here are not subject to flavour constraints, but mainly testable via precision measurements as well as direct searches for resonances. For all scenarios, the continuation of a strong experimental program at colliders is indispensable.

Acknowledgements I thank the organizers of the workshop for additional financial support, as well as A. Papaefstathiou and G. White for fruitful collaboration.

References

1. T. Robens, PoS CORFU2021, 031 (2022)
2. G.M. Pruna, T. Robens, Phys. Rev. **D88**, 115012 (2013), 1303.1150
3. T. Robens, T. Stefaniak, Eur. Phys. J. **C75**, 104 (2015), 1501.02234
4. T. Robens, T. Stefaniak, Eur. Phys. J. **C76**, 268 (2016), 1601.07880
5. A. Ilnicka, T. Robens, T. Stefaniak, Mod. Phys. Lett. A **33**, 1830007 (2018), 1803.03594
6. J. Alison et al., (2019), 1910.00012, [Rev. Phys.5,100045(2020)]
7. T. Robens, More doublets and singlets, in *56th Rencontres de Moriond on Electroweak Interactions and Unified Theories* (2022), 2205.06295
8. T. Robens, Di-Higgs production in BSM models, in *10th Large Hadron Collider Physics Conference* (2022), 2209.06795
9. T. Robens, T. Stefaniak, J. Wittbrodt, Eur. Phys. J. C **80**, 151 (2020), 1908.08554
10. T. Robens, Symmetry **15**, 27 (2023) 2209.10996
11. G. Altarelli, R. Barbieri, Phys. Lett. B **253**, 161 (1991)
12. M.E. Peskin, T. Takeuchi, Phys. Rev. Lett. **65**, 964 (1990)
13. M.E. Peskin, T. Takeuchi, Phys. Rev. D **46**, 381 (1992)
14. P. Bechtle, O. Brein, S. Heinemeyer, G. Weiglein, K.E. Williams, Comput. Phys. Commun. **181**, 138 (2010), 0811.4169
15. P. Bechtle, O. Brein, S. Heinemeyer, G. Weiglein, K.E. Williams, Comput. Phys. Commun. **182**, 2605 (2011), 1102.1898
16. P. Bechtle et al., Eur. Phys. J. C **74**, 2693 (2014), 1311.0055
17. P. Bechtle et al., Eur. Phys. J. **C80**, 1211 (2020), 2006.06007
18. P. Bechtle, S. Heinemeyer, O. Stål, T. Stefaniak, G. Weiglein, Eur. Phys. J. C **74**, 2711 (2014), 1305.1933
19. P. Bechtle et al., Eur. Phys. J. C **81**, 145 (2021), 2012.09197
20. Gfitter Group, M. Baak et al., Eur. Phys. J. **C74**, 3046 (2014), 1407.3792
21. J. Haller et al., Eur. Phys. J. **C78**, 675 (2018), 1803.01853
22. J. Alwall, M. Herquet, F. Maltoni, O. Mattelaer, T. Stelzer, JHEP **06**, 128 (2011), 1106.0522
23. LHC Higgs Cross Section Working Group, D. de Florian et al., (2016), 1610.07922
24. ATLAS, G. Aad et al., Phys. Lett. **B800**, 135103 (2020), 1906.02025
25. ATLAS, G. Aad et al., Phys. Rev. D **105**, 092002 (2022), 2202.07288
26. ATLAS, G. Aad et al., (2021), Phys Rev D **106**, 05200 (2022), 2112.11876
27. CERN Report No., (2021) (unpublished), ATLAS-CONF-2021-030
28. H. Abouabid et al., JHEP **09**, 011 (2022), 2112.12515
29. CERN Report No., (2013) (unpublished), CMS-PAS-HIG-13-003
30. CMS, V. Khachatryan et al., JHEP **10**, 144 (2015), 1504.00936
31. CMS, A.M. Sirunyan et al., JHEP **06**, 127 (2018), 1804.01939, [Erratum: JHEP03,128(2019)]
32. ATLAS, M. Aaboud et al., Phys. Rev. **D98**, 052008 (2018), 1808.02380
33. CERN Report No., (2012) (unpublished), CMS-PAS-HIG-12-045
34. M. Carena et al., Probing the electroweak phase transition with exotic higgs decays, in *2022 Snowmass Summer Study* (2022), 2203.08206
35. A. Papaefstathiou, G. White, JHEP **05**, 099 (2021), 2010.00597
36. A. Papaefstathiou, G. White, JHEP **02**, 185 (2022), 2108.11394
37. A. Papaefstathiou, T. Robens, G. White, Signal strength and W-boson mass measurements as a probe of the electro-weak phase transition at colliders—Snowmass White Paper, in *2022 Snowmass Summer Study* (2022), 2205.14379
38. Particle Data Group, P.A. Zyla et al., PTEP **2020**, 083C01 (2020)
39. Particle Data Group, R.L. Workman, PTEP **2022**, 083C01 (2022)
40. CERN Report No., 2021 (unpublished), ATLAS-CONF-2021-053
41. ATLAS, Nature **607**, 52 (2022), 2207.00092
42. CMS, Nature **607**, 60 (2022), 2207.00043

43. C. Vernieri, Higgs & BSM contributions, Talk at Energy Frontier Workshop, https://indico.fnal.gov/event/52465/contributions/236210/attachments/153456/199133/Snowmass-EF01-2-Brown.pdf
44. S. Dawson et al., Report of the topical group on higgs physics for snowmass 2021: the case for precision higgs physics, in *2022 Snowmass Summer Study*, 2022, 2209.07510
45. D. López-Val and T. Robens, Phys. Rev. D **90**, 114018 (2014), 1406.1043
46. M. Awramik, M. Czakon, A. Freitas, G. Weiglein, Phys. Rev. D **69**, 053006 (2004). (hep-ph/0311148)
47. A. Keshavarzi, D. Nomura, T. Teubner, Phys. Rev. D **101**, 014029 (2020), 1911.00367
48. T. Robens, TRSM Benchmark Planes—Snowmass White Paper, in *2022 Snowmass Summer Study* (2022), 2205.14486
49. T. Robens, Universe **8**, 286 (2022), 2205.09687
50. R. Coimbra, M.O.P. Sampaio, R. Santos, Eur. Phys. J. C **73**, 2428 (2013), 1301.2599
51. M. Mühlleitner, M.O.P. Sampaio, R. Santos, J. Wittbrodt, Eur. Phys. J. C **82**, 198 (2022), 2007.02985
52. P. Drechsel, G. Moortgat-Pick, G. Weiglein, Eur. Phys. J. C **80**, 922 (2020), 1801.09662
53. Y. Wang, M. Berggren, J. List (2020), 2005.06265, ILD-PHYS-PUB-2019-011
54. T. Robens, A short overview on low mass scalars at future lepton colliders—Snowmass White Paper, in *2022 Snowmass Summer Study* (2022), 2203.08210
55. CMS, Phys. Lett. B **842**, 137392 (2023), 2204.12413
56. CERN Report No., 2022 (unpublished), CMS-PAS-HIG-21-011
57. T. Robens, $b\bar{b}b\bar{b}$ final states in the TRSM for asymmetric production and decay. https://twiki.cern.ch/twiki/pub/LHCPhysics/LHCHWG3EX/rep.pdf
58. T. Robens, trsm_bbgaga.txt, https://twiki.cern.ch/twiki/bin/view/LHCPhysics/LHCHWG3EX
59. U. Ellwanger, C. Hugonie, Eur. Phys. J. C **82**, 406 (2022), 2203.05049
60. CMS, A. Tumasyan et al., Phys. Lett. B **835**, 137566 (2022), 2203.00480
61. ATLAS, G. Aad et al., Phys. Rev. D **105**, 012006 (2022), 2110.00313

Chapter 14
Searching for Light Physics at the LHC

Patrick Foldenauer

Abstract Over the last years, new physics in terms of a novel weakly-interacting massive particle (WIMP) has come more and more under pressure from experimental null results. While the remaining WIMP parameter space will be probed by next generation dark matter experiments, models of light new physics have become increasingly popular over the last decade. In an effort to explore the parameter space of such light physics, a myriad of custom designed high-precision/low-energy experiments has been proposed. In this note, however, I argue that existing LHC multipurpose experiments like ATLAS and CMS have a so far unexploited potential to probe light physics via appearing displaced recoil jets. In the first part, I discuss the sensitivity of this signature to (ultra-)light scalar and axionic dark matter, while in the second part I show its sensitivity to high-energy neutrino scattering.

14.1 Introduction

This note is a write-up of a talk given by the author at the 'Eighth Workshop on Theory, Phenomenology and Experiments in Flavour Physics (FPCapri2022)' in Villa Orlandi, Capri Island on June 17th 2022. The majority of the material presented here is based on work done in Refs. [1, 2] *in collaboration with Martin Bauer, Felix Kling, Peter Reimitz and Tilman Plehn.*

There is strong experimental evidence for physics beyond the Standard Model (SM). Some of the most compelling hints are for example the observation of neutrino oscillations [3], requiring massive neutrinos, or the latest Planck measurement of the CMB power spectrum, hinting at an overall dark matter (DM) abundance of $\Omega h^2 \sim 0.26$ [4]. Pinning down the correct particle physics model responsible for giving mass to the neutrinos and explaining the effect of dark matter with the correct relic

P. Foldenauer (✉)
Instituto de Física Teórica UAM-CSIC, Universidad Autónoma de Madrid,
Calle de Nicolás Cabrera 13-15, Cantoblanco, 28049 Madrid, Spain
e-mail: patrick.foldenauer@csic.es

G. Ricciardi et al. (eds.), *8th Workshop on Theory, Phenomenology and Experiments in Flavour Physics*, Springer Proceedings in Physics 292,
https://doi.org/10.1007/978-3-031-30459-0_14

abundance has been the goal of many particle physicists over the past decades. Many dedicated experiments have since been built to test interactions of both neutrinos and dark matter in order to gain a better understanding of their enigmatic nature. In this note, however, we want to focus on what complementary information we might gain from studying their interactions at LHC multipurpose experiments like ATLAS or CMS. To this purpose we will investigate the sensitivity of these experiments to ultralight dark matter candidates via appearing displaced jet signature in Sect. 14.2. In Sect. 14.3 we will then argue that a similar search strategy can also be exploited to observe interactions of the copiously produced neutrinos at the LHC.

14.2 Ultralight Dark Matter

Over the last decades the most popular model for DM has been in terms of a weakly-interacting massive particle (WIMP). This has been exhaustively searched for in collider as well as in direct and indirect detection experiments. While they still remain viable DM candidates, WIMP models have come more and more under pressure from experimental null results [5] and the remaining available parameter space will be probed by the next generation of DM experiments [6]. Below the GeV-scale WIMP-window there are many theoretically well-motivated light dark matter candidates, which can have relic abundances in agreement with observation from production via freeze-in, particle-anti-particle asymmetry, strong dynamics or non-thermal processes like misalignment [7]. In this section we want to focus on how to search for such light DM candidates.

14.2.1 Searches for Ultralight Dark Matter

It has been noted that for a particular mass scale of $m \sim 10^{-22}$ eV for the DM candidate the quantum pressure of the DM bose gas counteracts the gravitational collapse and helps to stabilise galaxy-sized DM halos, leading to a better fit of the small-scale matter power spectrum [8]. This has lead to increased interest in so-called ultra-light DM (ULDM) models over the last years.

However, this picture becomes more subtle as we also considers possible self-interactions of the ULDM particles. In case that the ULDM particles are subject to attractive self interactions, $\lambda < 0$, these will lead to instabilities of the dark matter halo and ultimately result in its collapse and the formation of boson stars [9]. On the contrary, if the self-interactions are repulsive ($\lambda > 0$) these will help to counteract the gravitational collapse of the halo and hence allow for a relaxed mass range for ULDM of $10^{-22} - 1$ eV [10]. These two scenarios are sketched in Fig. 14.1.

However, from a theoretical point of view it turns out to be rather difficult to construct well-motivated models for ULDM which naturally lead to repulsive self-

Fig. 14.1 Sketch of DM halo surrounding baryonic matter in galaxy. (Left) Repulsive self-interaction helps to counteract gravity and stabilises halo. (Right) Attractive self-interaction counteracts quantum pressure and destabilises halo

interactions [11]. Therefore, in a phenomenological approach in this note we will consider two simple extensions of the SM with either a new scalar or pseudo-scalar DM candidate.

The most straight-forward model to consider is that of a real scalar s coupled to the SM via the Higgs portal

$$\mathcal{L} \supset -\frac{1}{2}\lambda_{hs}\, s^2\, H^\dagger H\,. \tag{14.1}$$

At low energies, where we can integrate out the Higgs, this interaction leads to an effective DM-nucleon coupling of the form

$$\mathcal{L} \supset c_{sNN}\, s^2 \bar{N} N\,, \tag{14.2}$$

with a coupling coefficient

$$c_{sNN} = \lambda_{hs} \frac{m_N}{m_h^2} \frac{2n_H}{3(11 - \frac{2}{3}n_L)}\,. \tag{14.3}$$

This type of coupling leads to a potential between nuclei via the exchange of pairs of very light scalars s. Hence, it induces a fifth force that can be tested in precision measurements of low-energy observables in e.g. neutron scattering, molecular spectroscopy or Eot-Wash experiments [12]. Furthermore, since cold light DM has extremely high occupation numbers, it can be treated as a classical, coherently oscillating background field, which allows to separate the quadratic scalar interaction into a constant and a time-dependent part of the form

$$s^2 = s_0^2 \cos^2(m_s t) \to \frac{s_0^2}{2}\left(1 + \cos(2m_s t)\right)\,. \tag{14.4}$$

Ultimately, these terms will induce variations of fundamental constants [13, 14], or modifications of the primordial helium yield at the time of BBN [1].

For the simple Higgs portal model these constraints are collected for ultralight scalars in the left panel of Fig. 14.2. At higher masses such scalar DM candidates are

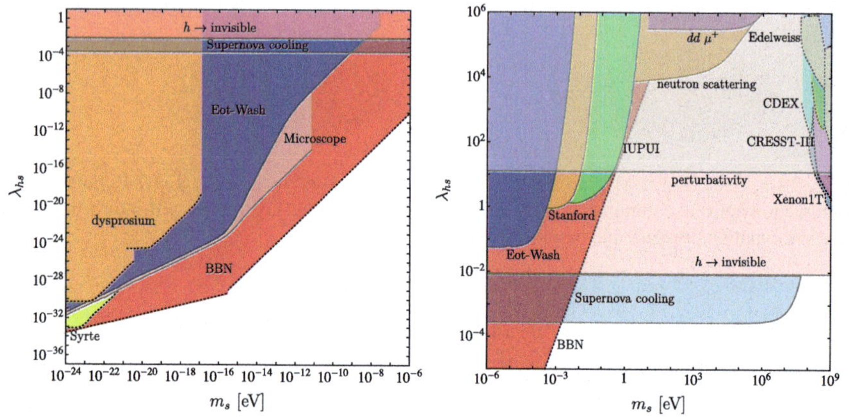

Fig. 14.2 Limits on a new scalar s coupled to the SM via the Higgs portal. Figures taken from [1]

mostly constrained by supernova energy loss, DM direct detection experiments [1] and the search for Higgs to invisible decays at LHC [15]. Limits in the high mass regime are collected in the right panel of Fig. 14.2.

14.2.2 Ultralight Dark Matter at the LHC

In this work we want to evaluate the potential of directly testing interactions of ULDM candidates at the LHC. To this purpose we are studying the sensitivity of LHC multipurpose experiments to two potential detection processes: *(i)* DM-background scattering and *(ii)* DM-nucleus scattering.

(i) DM-background scattering Staying with the case of the simple scalar Higgs portal model, we can produce the ULDM candidate s at the LHC from Higgs decays. The very boosted DM particle could then undergo scattering with the DM background field in the detector and e.g. produce a pair of visible photons via its Higgs interactions. This process is illustrated in the left panel of Fig. 14.3.

We can compute the mean free path $\lambda = 1/n_{\rm DM}\,\sigma$ for this type of process to get an estimate on which length scales these type of interactions are taking place. Assuming the maximum allowed coupling from Higgs to invisible decays $\lambda_{hs} = 8.7 \cdot 10^{-3}$, we arrive at a mean free path for this process of [1]

$$\lambda = \frac{4\pi}{\lambda_{hs}^2 g_{h\gamma\gamma}^2} \frac{m_h^3}{\rho_{\rm DM}} \gtrsim 10^{43}\ \mathrm{m}\,. \tag{14.5}$$

This is larger than the size of the observable universe by a factor of 10^{16}! Hence, this process is entirely unobservable and can be discarded.

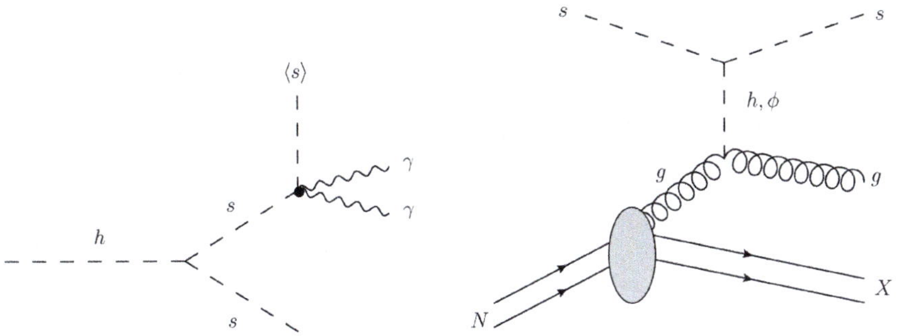

Fig. 14.3 (Left) Higgs decay into two ULDM scalars s with subsequent scattering of the boosted scalar with the DM background. (Right) ULDM-nucleus scattering of a real scalar s with the gluons via the exchange of a Higgs h (or new scalar mediator ϕ)

Fig. 14.4 Appearing displaced jet signature in an ATLAS-like detector from scattering of an ULDM particle. Figure taken from [1]

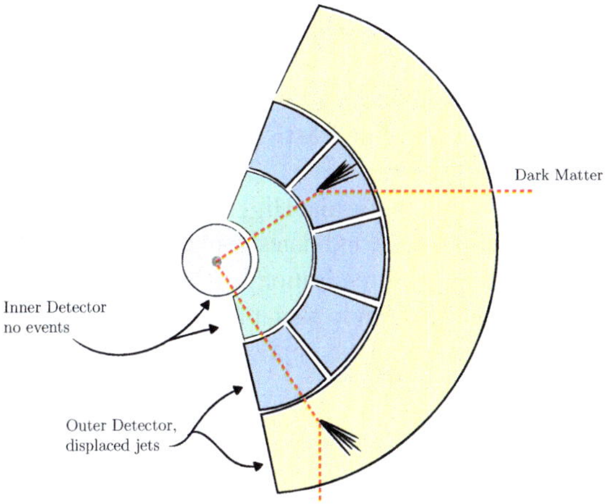

(ii) DM-nucleus scattering Alternatively, we can consider the same production mechanism of the ULDM particles via Higgs decay, but now consider scattering with the detector material itself via the Higgs portal interaction. This process is depicted in the right panel of Fig. 14.3.

We can compute the interaction probability of the DM interacting in a detector of length L_X and nuclear density n_X via the mean free path $\lambda_X = 1/n_X\,\sigma_X$,

$$P_{\text{int}} = 1 - e^{-L_X/\lambda_X} = 1 - e^{-L_X\, n_X\, \sigma_X}\ . \tag{14.6}$$

Hence, in this process the interaction probability crucially depends on the density of the detector material. In Fig. 14.4 we show the typical interaction signature at an ATLAS-like detector. Since the inner parts of the detector are gaseous xenon trackers, the material density n_{Xe} and hence the interaction probability is very low. However, the outer calorimeter parts are mainly made of lead and iron with much higher densities, $n_{Pb} \gg n_{Xe}$. Thus, the DM particle can scatter in these outer, denser detector components and produce a displaced recoil jet [1]. The partonic DM-nucleus scattering cross section is typically expressed via the energy loss ν of the DM particle and the kinematic variables $x = Q^2/(2M\nu)$ and $y = \nu/E_s$,

$$\frac{d^2\hat{\sigma}_{\text{DIS}}}{dx\,dy} = \frac{\lambda_{hs}^2 g_{hgg}^2}{4\pi\,\hat{s}} \frac{Q^4}{(Q^2 + m_h^2)^2}\,, \tag{14.7}$$

where $\hat{s} = xs = 2ME_s\,x$ and $Q^2 = 2ME_s\,x\,y$. By means of this cross section we can compute the interaction probability in an ATLAS-like detector as

$$P_{\text{DIS}} = 1 - e^{-L_{\text{E}}\,n_{\text{Pb}}\,\sigma_{\text{Pb}}} e^{-L_{\text{H}}\,n_{\text{Fe}}\,\sigma_{\text{Fe}}} \approx 7.5 \cdot 10^{-21}\,. \tag{14.8}$$

Hence, for the simple Higgs portal interaction this process is again unobservable even at the HL-LHC which aims to collect $\mathcal{O}(10^8)$ Higgses.

14.2.2.1 Axion-like Particles

Beyond the minimal scalar Higgs portal model, axions and axion-lke particles (ALPs) are well-motivated candidates to play the role of ULDM. Therefore, as an alternative phenomenological realisation of ULDM we consider an extension of the SM by a shift-symmetric pseudo-scalar (or axion-like) particle a. Such a particle can arise as the angular mode of a complex scalar S acquiring a vacuum expectation value f,

$$S = \frac{s + f}{\sqrt{2}}\, e^{ia/f}\,. \tag{14.9}$$

This angular mode a is protected by a shift symmetry of the type

$$e^{ia/f} \to e^{i(a+c)/f} = e^{ia/f} e^{ic/f}\,, \tag{14.10}$$

by a constant shift c.

If these particles are coupled to QCD they will acquire a mass term by an explicit breaking of the shift symmetry via non-perturbative effects leading to a potential for the ALP of

$$V(a) = \Lambda^4 \left[1 - \cos\left(\frac{a}{f}\right)\right] \simeq \frac{\Lambda^4}{2f^2}\, a^2 + \ldots\,. \tag{14.11}$$

Hence, this potential generates a mass for the ALP of $m_a = \Lambda^4/2f^2$, where Λ is the QCD scale and f the heavy axion scale. Thus, ALPs can quite generically have very suppressed masses, making them good candidates for ULDM.

As a toy model for ultralight ALPs, we consider here a shift-symmetric, pseudo-scalar particle a coupled to the SM via a new heavy scalar mediator ϕ via the interaction,

$$\mathcal{L} \supset \frac{1}{2}\partial_\mu a\, \partial^\mu a - \frac{m_a^2}{2}a^2 + \frac{1}{2}\partial_\mu \phi\, \partial^\mu \phi - \frac{1}{2}m_\phi^2 \phi^2 - \frac{\partial_\mu a\, \partial^\mu a}{2\Lambda_{\phi a}}\phi - \frac{\alpha_s}{\Lambda_\phi}\,\phi\, \mathrm{Tr}\,[G_{\mu\nu}G^{\mu\nu}]\,. \tag{14.12}$$

Similar to the simple Higgs portal model, the interactions of Eq. 14.12 lead to an effective low-energy ALP-nucleon coupling,

$$\mathcal{L} \supset c_{aNN}\, \partial_\mu a\, \partial^\mu a\, \bar{N}N\,, \tag{14.13}$$

with an effective coupling coefficient of [1]

$$c_{aNN} = \frac{m_N}{\Lambda_{\phi a}\Lambda_\phi m_\phi^2}\frac{8\pi}{11 - \frac{2}{3}n_L}\,. \tag{14.14}$$

The major difference in the phenomenology of these pseudo-scalar ULDM candidates to scalars is the appearance of the derivative coupling in Eq. 14.13. This leads to a quadratic scaling of the effective ALP-nucleon coupling with the momentum transfer q of the interaction. Since all the high-precision searches for ULDM considered in Sect. 14.2.1 are performed at very low energies (compared to the typical weak-scale suppression scales Λ_ϕ, $\Lambda_{\phi a}$) the effective ALP interactions are heavily suppressed compared to the simple real scalar Higgs portal model. This is reflected in the large unconstrained region of parameter space in the left panel of Fig. 14.5.

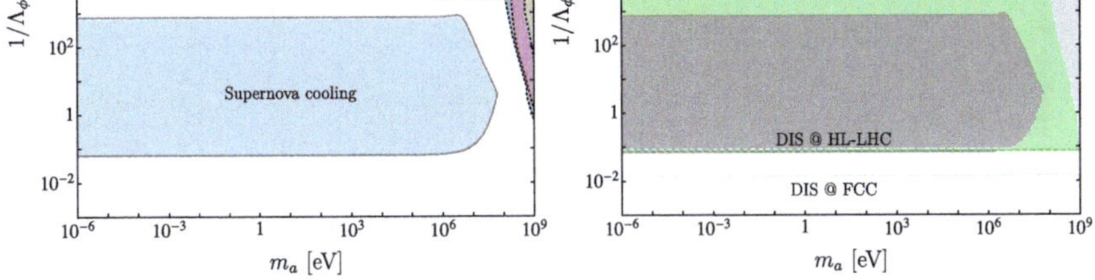

Fig. 14.5 Limits on a new pseudo-scalar a coupled to the SM via a new heavy mediator ϕ (left). Appearing displaced jet search sensitivity at HL-LHC and FCC (right). Figures taken from [1]

These ultralight ALPs can be produced in hadron colliders via the decay of the heavy mediator ϕ, which is produced via its coupling to gluons in Eq. 14.13. These ALPs a can scatter in the detector material via the same process we considered in the case of the real scalar s. Hence, we can compute the interaction probability of a with the detector via Eq. 14.6 using the relevant DIS cross section

$$\frac{d^2\hat{\sigma}_{\rm DIS}}{dx\,dy} = \frac{\alpha_s^2}{16\pi\,\hat{s}}\,\frac{Q^4}{\Lambda_{\phi a}^2 \Lambda_{\phi^2}} \left(\frac{Q^2 + 2m_a^2}{Q^2 + m_\phi^2}\right)^2 . \tag{14.15}$$

The sensitivities of ATALS at high-luminosity LHC and a future FCC to the resulting displaced jet signature of the pseudo-scalar ULDM candidate is shown in the right panel of Fig. 14.5. It can be seen that the LHC search will be able to probe a large part of the currently allowed parameter space and is competitive with bounds from supernova cooling. In the future, FCC will be able to probe couplings which are smaller by even an order of magnitude. The reason why this displaced jet search at high-energy colliders has such a high sensitivity to this model is the derivative interaction of the ALPs a. While this leads to a suppression of the interaction rates at low-energy experiments, it results in an enhancement of the signal at colliders.

14.3 Neutrinos

In the second part of this work we will investigate the sensitivity of a displaced jet search to scattering of SM neutrinos in LHC multipurpose detectors.

While neutrino scattering has been studied in a number of experiments over the last decades, there is still only very limited data available for high-energy interactions. In Fig. 14.6 one can see a compilation of neutrino scattering cross section measurements for the different neutrino flavours together with the expected neutrino spectra at the future LHC forward experiment FASERν. As can be seen we have

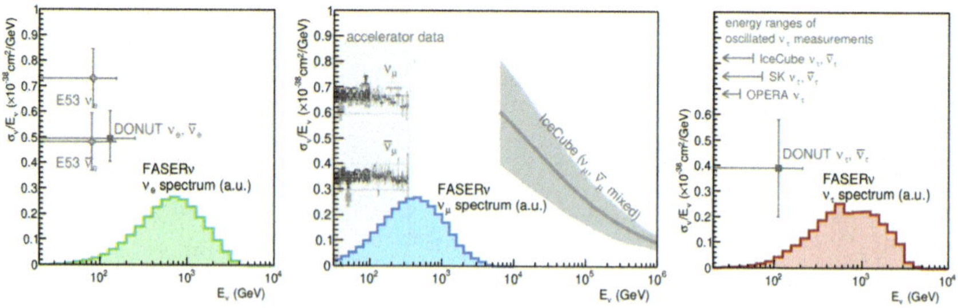

Fig. 14.6 Past neutrino cross section measurements for electron- (left), muon- (centre) and tau-flavoured neutrinos alongside the expected neutrino spectra at FASERν. Figure taken from [16]

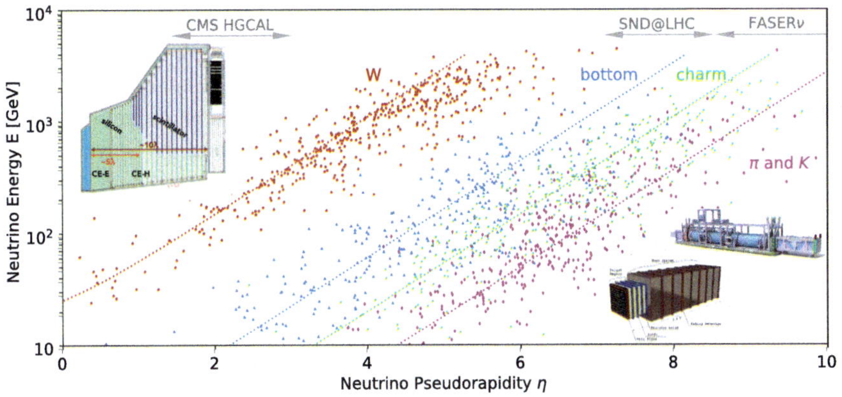

Fig. 14.7 Sensitivities of various LHC based experiments to the LHC neutrino flux. Sources considered are neutrinos from W boson, bottom, charm as well as π and K meson decays. Figure taken from [2]

by far most data for ν_μ-scattering. However, at intermediate energies above accelerator experiments and below the high-energy events at IceCube [17] there still is a blind spot to ν_μ-scattering at energies between 370 GeV and 6.3 TeV. For e- and τ-flavoured neutrinos there is still no data above $\mathcal{O}(100)$ GeV energies at all.

In Fig. 14.7 we show the energy distribution of the LHC produced neutrino flux as a function of pseudorapidity η for a number of sources. In this figure we illustrate the angular sensitivities of the future forward experiments SND@LHC and FASERν alongside the upgraded CMS endcap high-granularity calorimeter (HGCAL). As can be seen, the forward physics experiments are excellently suited to detect a large fraction of the flux of high-energy neutrinos from meson decays. However, there is still a large unexploited reservoir of neutrinos produced from W decays in the process $W \to \nu_\mu \, \mu$, which is in the angular acceptance of the CMS HGCAL. This serves as a strong motivation to investigate the HGCAL sensitivity to detect these neutrinos via scattering.

14.3.1 How to Detect Neutrinos at CMS

Inspired by our study of detecting ULDM scattering via appearing displaced jet signatures at LHC multipurpose experiments, we are investigating appearing neutral jets from neutrino-nucleus scattering in the calorimeter material as a detection signature. Isolating these neutrino scattering events at a hadron collider is extremely challenging due to the large background of neutral hadrons from underlying event and pile-up. However, the CMS HGCAL upgrade has some very advantageous features to tackle this background.

The HGCAL is a sampling detector with an angular coverage of $1.5 < |\eta| < 3.0$. It consists mainly of silicon as active material with a size of $0.5\,\mathrm{cm}^2$–$1\,\mathrm{cm}^2$ of the individual cells in the transverse plane [18]. Such a finely granulated detector enables a high-resolution measurement of the lateral development of electromagnetic showers. This is crucial for a good two-shower separation and the observation of narrow jets. Furthermore, the HGCAL has excellent timing capabilities and is able to operate with a timing window of 90 ps to remove pile-up, corresponding to a path length difference travelled by relativistic particles of $\Delta_l \sim 2.7$ cm. A further step to remove pile-up even more in the future is a novel proposed collision technique for the HL-LHC, the so-called *crab kissing* [19]. Here, the bunches are colliding at a crossing angle. In order to achieve maximum spatial overlap, they pass a radio frequency cavity giving the tail and head of the bunches a kick so that they rotate. With a typical spatial extension of 31.4 cm of the bunch and about 130 expected pile-up events per bunch crossing [20], the timing resolution will allow to reduce the number of pile-up events per bunch crossing on average to

$$N_{\mathrm{pu}} \approx \frac{2.7}{31.4}\,130 \sim 11\,. \tag{14.16}$$

The neutrino signal which we are focusing on is given by ν_μ produced in a W decay in conjunction with a charged muon, and the subsequent scattering of the neutrino with the detector. The scattering of the ν_μ with the detector occurs mostly via the charged current interaction at high energies. There the neutrino converts into a charged lepton and the scattered quark will confine into a jet. Hence, the resulting characteristic signature is a charged lepton together with a single energetic jet produced in a single displaced vertex in the calorimeter with no tracks pointing back to the interaction point.

The signal is hence characterised by the process,

$$\text{Production:}\quad qq' \to W \to \mu_1\,\nu_\mu\,, \tag{14.17}$$

$$\text{Scattering:}\quad \nu_\mu N \to \mu_2 + \text{jet}\,, \tag{14.18}$$

where a prompt primary muon μ_1 originates from production and a secondary displaced muon μ_2 appears in the scattering. This signal topology in the CMS HGCAL is illustrated in the left panel of Fig. 14.8. To determine the neutrino scattering rate with the detector, we have used the deep-inelastic neutrino interaction cross-sections at leading order as derived in Ref. [21].

The leading backgrounds to this signal are isolated muons produced in conjunction with a long-lived neutral hadron interacting in the calorimeter. Potential candidates to mimic the displaced jet of a neutrino signal are neutrons, K_L^0, Λ, or Ξ^0. These hadrons are produced copiously in hard scattering, the underlying event or originate from pile-up. Since these are very likely to interact in the hadronic calorimeter our main background consists of

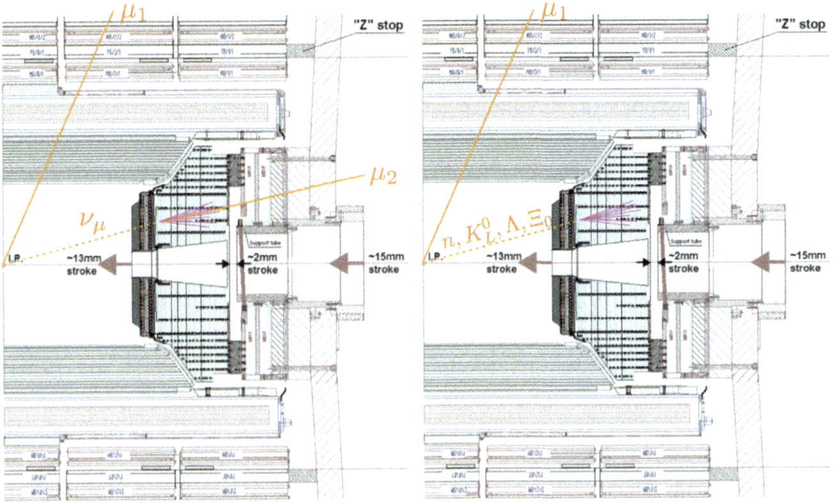

Fig. 14.8 Signal (left) and background (right) topology for neutrino scattering in the CMS detector leading to an appearing jet in the HGCAL

(i) The decay of a W-boson into leptons in association with additional neutral hadrons:

$$qq' \to W + \text{QCD} \to \mu_1 + \text{QCD}, \tag{14.19}$$

(ii) Heavy quark production with additional neutral hadrons, with the heavy flavour decaying into muons:

$$qq' \to b/c + \text{QCD} \to \mu_1 + \text{QCD}. \tag{14.20}$$

In these processes an appearing jet is produced form the scattering of the neutral hadron in the calorimeter:

$$\text{neutral hadron} + N \to \text{jet}. \tag{14.21}$$

The corresponding background topology in the CMS HGCAL is shown in the right panel of Fig. 14.8. As can be seen immediately a striking discriminating feature is the secondary muon in the signal.

In our study we have simulated both signal and background for a data set of 3 ab^{-1} at leading order including parton shower and hadronisation with Pythia 8.2 [22]. Our analysis strategy is then as follows:

a. *Isolated central muon*: We start by requiring a central, isolated primary muon with

$$R_{\text{iso},\mu_1} > 0.1, \quad p_{T,\mu_1} > 20\ \text{GeV}, \quad |\eta_{\mu_1}| < 2.4\,. \tag{14.22}$$

b. *Isolated jet*: To reject the fake background jets we use that energetic neutral hadrons are mostly produced as part of an energetic, and hence collimated, hadronic shower. To distinguish between jets from neutral hadrons and neutrinos, we therefore apply a jet isolation cut,

$$R_{\text{iso},j} > 0.1\,. \tag{14.23}$$

c. *W mass cut*: Since our signal is consisting of a muon and a muon-neutrino originating from a W decay, we require the neutrino jet and the primary muon to reconstruct the W mass. Defining $(p_{\mu_1} + p_\nu)^2 = m_{\mu\nu}^2$ we hence require

$$66\text{ GeV} < m_{\mu\nu} < 99\text{ GeV}\,. \tag{14.24}$$

d. *Secondary muon*: We can search for the additional secondary muon as an extra handle to isolate the signal with the CMS endcap muon stations located behind the HGCAL covering $1.2 < |\eta| < 2.4$. Hence, we require the hardest muon originating from the detector material collision to carry away

$$E_{\mu_2}/E_\text{j} > 0.33\,. \tag{14.25}$$

e. *Energy cut*: Since neutrinos produced in W boson decays are expected to be more energetic than the neutral hadron background, we only consider events with displaced jet energies of

$$E_\text{cut} > 160\,\text{GeV}\,. \tag{14.26}$$

In Fig. 14.9 we show the signal versus background distributions both in the jet energy versus R and muon-jet invariant mass plane. For illustration we also display the jet-isolation, the invariant mass reconstruction and the energy cut. As can be seen these cuts are well suited to separate the signal from the background. However, the most important selection criterion is the detection of the energetic secondary muon, which can suppress the background by almost four orders of magnitude. The full cut flow table is shown on the left of Fig. 14.10, which shows that this analysis is able to reduce the initial huge hadron background to a similar level of remaining neutrino signal events. In the right panel of Fig. 14.10 we show the final histogram of signal and background events as a function of jet energy. In summary, we can see that in terms of a proof-of-principle strategy this analysis is capable of detecting neutrino scattering at the CMS HGCAL.

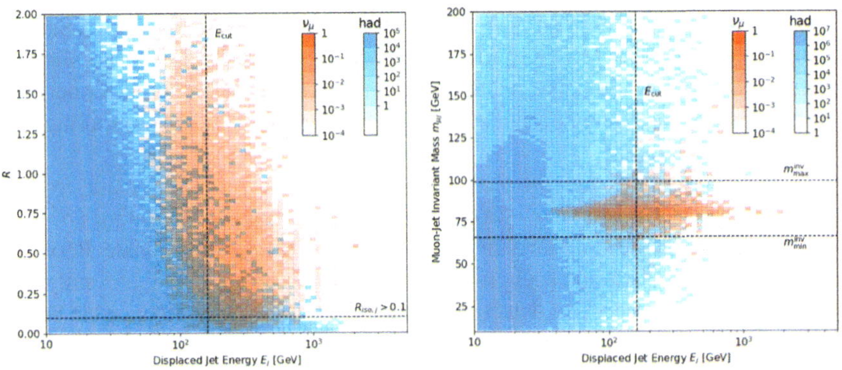

Fig. 14.9 Distribution of signal (red) and background (blue) events in the jet energy versus R plane (left) and in the jet energy versus muon-jet invariant mass plane (right). The left histogram shows the selection after applying the isolated muon cut and the right one after applying the jet isolation cut. Figures taken from [2]

Cuts	Hadrons	Neutrinos
isolated muon	$1.02 \cdot 10^{11}$	7.59
isolated jet	$8.63 \cdot 10^{10}$	7.05
W mass	$1.92 \cdot 10^{9}$	6.55
secondary muon	$3.49 \cdot 10^{5}$	5.48
$E_j > 160$ GeV	3.52	3.60

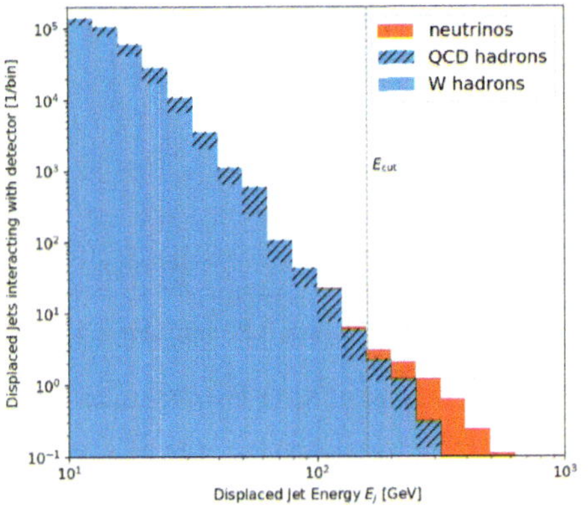

Fig. 14.10 (Left) Analysis cut flow table. (Right) Final histogram of signal (red) and background (red) events after all cuts. Taken from [2]

14.4 Conclusions

Summarising our results presented in the last two sections, we can conclude that appearing displaced jets are a promising signature to search for the scattering of light elusive particles in LHC multipurpose experiments.

In particular, in Sect. 14.2 we have argued that high-energy appearing jets in the ATLAS calorimeters are a sensitive probe of large parts of the parameter space of a shift-symmetric light pseudo-scalar particle a coupled to the SM via a new heavy mediator. This is mainly due to the momentum-enhanced scattering cross section at high-energy colliders over low-energy precision probes.

In Sect. 14.3 we presented a proof-of-principle analysis of appearing neutrino-scattering jets at the CMS HGCAL. The crucial handle on the neutrino signal to discriminate it from the large neutral hadronic background is the detection of an energetic secondary muon produced in the charge current neutrino scattering with the detector material.

In the future, there are still many possible improvements to be made to these search strategies. In particular, one can apply machine learning techniques to study the lateral shower development in the detectors to better discriminate BSM and neutrino jets from hadronic background jets. For neutrino scattering, it is still worth investigating the prospects of detecting scattering from neutrinos produced in bottom and charm flavoured meson decays. Furthermore, one could search for neutrino scattering directly in the central muon stations.

Acknowledgements The author would like to express special thanks to the Mainz Institute for Theoretical Physics (MITP) of the Cluster of Excellence PRISMA+ (Project ID 39083149), for its hospitality and support. The work of PF was partially supported by the UKRI Future Leaders Fellowship DARKMAP and the Spanish Agencia Estatal de Investigacíón through the grant CEX2020-001007-S, funded by MCIN/AEI/10.13039/501100011033.

References

1. M. Bauer, P. Foldenauer, P. Reimitz, T. Plehn, Sci. Post Phys. **10**, no. 2, 030 (2021). [arXiv:2005.13551 [hep-ph]]
2. P. Foldenauer, F. Kling, P. Reimitz, Phys. Rev. D **104**, no. 11, 113005 (2021). [arXiv:2108.05370 [hep-ph]]
3. Q.R. Ahmad et al., SNO. Phys. Rev. Lett. **87**, 071301 (2001). ([arXiv:nucl-ex/0106015 [nucl-ex]])
4. N. Aghanim et al., [Planck], Astron. Astrophys. **641** (2020), A6 [erratum: Astron. Astrophys. **652** (2021), C4] [arXiv:1807.06209 [astro-ph.CO]]
5. R.K. Leane, T.R. Slatyer, J.F. Beacom, K.C.Y. Ng, Phys. Rev. D **98**, no. 2, 023016 (2018). [arXiv:1805.10305 [hep-ph]]
6. G. Arcadi, M. Dutra, P. Ghosh, M. Lindner, Y. Mambrini, M. Pierre, S. Profumo, F.S. Queiroz, Eur. Phys. J. C **78**, no.3, 203 (2018). [arXiv:1703.07364 [hep-ph]]
7. S. Knapen, T. Lin, K.M. Zurek, Phys. Rev. D **96**, no. 11, 115021 (2017).[arXiv:1709.07882 [hep-ph]]
8. W. Hu, R. Barkana, A. Gruzinov, Phys. Rev. Lett. **85**, 1158–1161 (2000). [arXiv:astro-ph/0003365 [astro-ph]]
9. A.H. Guth, M.P. Hertzberg, C. Prescod-Weinstein, Phys. Rev. D **92**, no. 10, 103513 (2015). [arXiv:1412.5930 [astro-ph.CO]]
10. E.G.M. Ferreira, Astron. Astrophys. Rev. **29**, no. 1, 7 (2021). [arXiv:2005.03254 [astro-ph.CO]]
11. J. Fan, Phys. Dark Univ. **14**, 84–94 (2016). ([arXiv:1603.06580 [hep-ph]])
12. P. Brax, S. Fichet, G. Pignol, Phys. Rev. D **97**, no. 11, 115034 (2018). [arXiv:1710.00850 [hep-ph]]
13. Y.V. Stadnik, V.V. Flambaum, Phys. Rev. Lett. **115**, no. 20, 201301 (2015). [arXiv:1503.08540 [astro-ph.CO]]
14. A. Hees, O. Minazzoli, E. Savalle, Y.V. Stadnik, P. Wolf, Phys. Rev. D **98**, no. 6, 064051 (2018). [arXiv:1807.04512 [gr-qc]]
15. C. Bernaciak, T. Plehn, P. Schichtel, J. Tattersall, Phys. Rev. D **91**, 035024 (2015). ([arXiv:1411.7699 [hep-ph]])

16. H. Abreu et al., [FASER]. [arXiv:2001.03073 [physics.ins-det]]
17. M.G. Aartsen et al., IceCube. Nature **551**, 596–600 (2017). ([arXiv:1711.08119 [hep-ex]])
18. [CMS], CERN-LHCC-2017-023
19. S. Fartoukh, Phys. Rev. ST Accel. Beams **17**, no. 11, 111001 (2014)
20. L.E. Medina Medrano, R. Tomas Garcia, G. Arduini, M. Napsuciale, CERN-ACC-2018-0003. https://cds.cern.ch/record/2301928
21. H. Abreu et al., [FASER], Eur. Phys. J. C **80**, no. 1, 61 (2020). [arXiv:1908.02310 [hep-ex]]
22. T. Sjöstrand, S. Ask, J.R. Christiansen, R. Corke, N. Desai, P. Ilten, S. Mrenna, S. Prestel, C.O. Rasmussen, P.Z. Skands, Comput. Phys. Commun. **191**, 159–177 (2015). ([arXiv:1410.3012 [hep-ph]])

The manufacturer's authorised representative in the EU is Springer Nature Customer Service Centre GmbH, Europaplatz 3, 69115 Heidelberg, Germany. If you have any concerns regarding our products, please contact ProductSafety@springernature.com

Printed and bound by CPI Group (UK) Ltd, Croydon, CR0 4YY
01/10/2026
02225881-0002